Mastering the Periodic Table: *50 Activities on the Elements*

Linda Trombley

Revised by Faye Williams

WALCH PUBLISHING

User's Guide to Walch's Reproducible Books

As part of our general effort to provide educational materials that are as practical and economical as possible, we have designated this publication a "reproducible book." The designation means that the purchase of the book includes purchase of the right to limited reproduction of all pages on which this symbol appears:

Here is the basic Walch policy: We grant to individual purchasers of this book the right to make sufficient copies of reproducible pages for use by all students of a single teacher. This permission is limited to a single teacher and does not apply to entire schools or school systems, so institutions purchasing the book should pass the permission on to a single teacher. Copying of the book or its parts for resale is prohibited.

Any questions regarding this policy or request to purchase further reproduction rights should be addressed to:

Permissions Editor
J. Weston Walch, Publisher
P.O. Box 658
Portland, Maine, 04104-0658

1 2 3 4 5 6 7 8 9 10
ISBN 0-8251-3937-6

J. Weston Walch, Publisher
P.O. Box 658 • Portland, Maine 04104-0658
www.walch.com
Printed in the United States of America

Contents

To the Teacher

This set of activities has been created and assembled to provide a variety of materials for both students and teacher. The activities are designed to cover every aspect of the periodic table and are divided into categories (listed in the Table of Contents). The types of activities range from worksheets to demonstrations and from group projects to games and puzzles. Some activities are cross-curricular and incorporate history (Activity 9 or 48), music (Activity 1), art (Activity 2 or 18), writing (Activity 1 or 31b), and math (Activity 13). You can select an activity to introduce an idea or to reinforce prior knowledge. The activities can be used singly or as a complete unit. Depending on the class, one activity can be assigned as remediation to clarify a concept (Activity 19, 29, or 30) and another activity can be given as an enrichment assignment (Activity 11, 28b, or 31b). With Activities 12–14, students make up their own stories or problems using the elements.

The reproducible readings are designed to provide the teacher with enough information to explain the design, uses, and complexities of the periodic table. They are also intended for students to read for background information prior to completing the selected activity. In fact, a simple recall activity, such as Activity 21a or 32a, can be done after reading and before an assignment. This same recall activity format can be used for element properties, electron configuration, or specific series or families of elements.

Depending on the ability level of the students, the activities can be used for middle school, junior high, or high school students. Some of the puzzles can be done by fourth or fifth graders using the reproducible readings as guides. Younger students can complete some of the quizzes as worksheets when used in conjunction with the reproducible readings.

For biology classes, Activities 3–8, 12–17, 32–34 and Bonus Activities 1 and 2 could be useful. For general science classes, Activities 1–9, 12–22, 32–34 and Bonus Activities 1 and 2 are suggested. Physical science teachers may want to use the same activities as suggested for general science, plus Activities 37–42. The majority of the activities are designed for chemistry classes. However, as indicated above, you should select the activities that are best suited to your classroom situation and students. Some of the activities can even be used as handouts for a substitute, especially Activities 12–14. All the substitute needs is a periodic table and the worksheet. The appendices also contain a short list of web sites that have links to other sources. Classrooms with Internet capability can open an even wider vista for students to explore.

As the introductory reading states: The periodic table of elements is the most important tool of a chemist. These activities are designed to help students discover the myriad of information available from this simple chart of elements and understand how this information affects their lives.

Introductory Information

Preliminary Handouts:

- About the Periodic Table (reproducible reading)
- Periodic Table of Elements (reproducible chart)

Activities:

1. Element Song (teacher guide page)
2. Element Boxes (teacher guide page)

About the Periodic Table

The **periodic table of elements** is the most important tool of a chemist. The table provides a myriad of information about all the elements, both natural and synthetic.

By the year 2000, there were **115 elements** — 92 occurring naturally and 23 synthetic. Technetium and all the transuranium elements are synthetic. Plutonium is found in very minute quantities in uranium ores, so it is counted as part of the 92 naturally occurring elements; however, the majority of plutonium is synthetic.

The elements have been placed in **periods** (the horizontal rows). These indicate electron energy levels. They are also placed in **groups** or **families** (the vertical columns); these reflect similar chemical properties. There are 7 periods and 18 groups, numbered from left to right. All elements from bismuth (atomic number 83) to the latest synthetic element are radioactive.

The types of information available from the periodic table vary depending on the version, but each table has the element symbol and the atomic number. Other types of information that can be found on the periodic table are:

- element name
- mass number
- atomic mass
- electron configuration
- oxidation states
- phase at room temperature
- melting point
- boiling point
- density
- atomic radius
- electronegativity

Chemists are able to use the periodic table to determine which elements can combine and in what proportions and which elements have similar properties. Using this information they can help manufacturers determine how to improve products and processes. For example, the substitution of potassium for sodium strengthens glass, and the use of helium instead of hydrogen in dirigibles provides safety.

The periodic table on page 3 is based on the existing published periodic tables and information gathered from various web sites. The published periodic tables are from the American Chemical Society (ACS). Most periodic table sources do not identify the latest elements. Gesellschaft für Schwerionenforschung (GSI), currently one of the leading researchers in developing new elements, provides the latest information on their web site. GSI follows International Union of Pure and applied Chemistry (IUPAC) guidelines. The periodic table found in the Webelements web site does list elements 113, 115, and 117 although these have not, as of 2000, been identified.

Periodic Table of Elements

Period	1	2	3	4	5	6	7	8	9	10	11	12	13	14	15	16	17	18
	1																	0
1	1 H																	2 He
	IA	IIA	IIIB	IVB	VB	VIB	VIIB	VIIIB			IB	IIB	IIIA	IVA	VA	VIA	VIIA	
2	3 Li	4 Be											5 B	6 C	7 N	8 O	9 F	10 Ne
3	11 Na	12 Mg											13 Al	14 Si	15 P	16 S	17 Cl	18 Ar
4	19 K	20 Ca	21 Sc	22 Ti	23 V	24 Cr	25 Mn	26 Fe	27 Co	28 Ni	29 Cu	30 Zn	31 Ga	32 Ge	33 As	34 Se	35 Br	36 Kr
5	37 Rb	38 Sr	39 Y	40 Zr	41 Nb	42 Mo	43 Tc	44 Ru	45 Rh	46 Pd	47 Ag	48 Cd	49 In	50 Sn	51 Sb	52 Te	53 I	54 Xe
6	55 Cs	56 Ba	57 La	72 Hf	73 Ta	74 W	75 Re	76 Os	77 Ir	78 Pt	79 Au	80 Hg	81 Tl	82 Pb	83 Bi	84 Po	85 At	86 Rn
7	87 Fr	88 Ra	89 Ac	104 Rf	105 Db	106 Sg	107 Bh	108 Hs	109 Mt	110 Uun	111 Uuu	112 Uub		114 Uuq		116 Uuh		118 Uuo

6 - Lanthanides	58 Ce	59 Pr	60 Nd	61 Pm	62 Sm	63 Eu	64 Gd	65 Tb	66 Dy	67 Ho	68 Er	69 Tm	70 Yb	71 Lu
7 - Actinides	90 Th	91 Pa	92 U	93 Np	94 Pu	95 Am	96 Cm	97 Bk	98 Cf	99 Es	100 Fm	101 Md	102 No	103 Lr

Represents period numbers

Represents new group numbers

Represents old group numbers

Activity 1: **Element Song**

Objective: Students creatively present information about an element.

Materials:
- Recording of "The Element Song" by Tom Lehrer
- Copy of the periodic table of elements for each student (can be reproduced on page 3)
- Resources for information on elements (for example, *CRC Handbook of Chemistry and Physics*)

Procedure: If possible, play "The Element Song" by Lehrer. Discuss with the class. Have students select an element and write a song or poem about it. They should be sure to include the element name, symbol, atomic number, group number or family name, major characteristics, and uses.

Results: Students will present their songs or poems to the class. This is an opportunity for students to be creative and have some fun with a potentially dry subject.

Variation:
1. Students can update Lehrer's song to include the latest elements.
2. Students can work in groups, with each group writing a song about the elements in an assigned period from the periodic table (horizontal row).

Activity 2: **Element Boxes**

Objective: Students create their own periodic table.

Materials:
- 15-cm x 15-cm squares of white paper
- Resources for information on elements (for example, *The Elements* by John Emsley)

Procedure: Have students each select an element and design an "element box" for it. The box (white paper square) should include element name, symbol, atomic number, and a picture representative of either the element's characteristics or uses.

Results: Students will place their element boxes in the correct periodic table location on a large bulletin board or piece of poster board.

Extension: As they continue to learn about the periodic table, students can add more information to their element boxes. For example, atomic mass or mass number, electronegativity, phase at room temperature, date of discovery, mineral source, oxidation states, etc., can be included as time goes on.

Metals, Semimetals, and Nonmetals

Preliminary Handout:

- Metals, Semimetals, and Nonmetals (reproducible reading)

Activities:

3. Identifying by Color (teacher guide page)
4. The Metals Crossword (reproducible)
5. Nonmetals Word Find (reproducible)
6. Mixed-Up Table (Semimetals) (reproducible)

Metals, Semimetals, and Nonmetals

The periodic table of elements can be used to determine metals, semimetals, and nonmetals. Metals make up the majority of the elements. They are located to the left of the semimetals on the periodic table. (**Hydrogen**, however, is a nonmetal.) Metals tend to lose electrons and form positive ions or **cations** in chemical reactions. The semimetals are those elements that have properties that are not distinctly metallic or nonmetallic. These elements are **boron**, **silicon**, **germanium**, **arsenic**, **antimony**, **tellurium**, and **astatine**. Hydrogen and the elements to the right of the semimetals are nonmetals. Nonmetals tend to gain electrons and form negative ions or **anions** in chemical reactions.

Depending on the element with which hydrogen is reacting, hydrogen will form a positive or a negative ion. When bonding with other nonmetals, hydrogen forms a positive ion, but when bonding with metals, hydrogen forms a negative ion.

Activity 3: **Identifying by Color**

Objective: Students will identify metals, semimetals, and nonmetals on the periodic table.

Materials:
- Copy of the periodic table for each student (can be reproduced on page 3)
- Colored pencils or crayons (30 each of yellow, blue, and green)

Procedure: Have students color their periodic tables to indicate metals, semimetals, and nonmetals. The semimetals should be a blend of the two colors used for metals and nonmetals. For example, metals can be colored yellow, nonmetals blue, and semimetals green (blue and yellow make green). This is to indicate that characteristics of semimetals are a blending of the characteristics of metals and nonmetals.

Results: Students can see how the periodic table is divided so metals are on the left side and nonmetals are on the right side, with the exception of hydrogen. They should be able to name the seven semimetals and associate the blending of the colors with the blending of both metal and nonmetal characteristics.

Extensions: Use the same periodic table for identifying groups and charges. For groups, students can add pictures to the element boxes. For example, alkali metals can be represented by a flame since they react with water to release flammable hydrogen. For charges, students can add + or – signs. For example, lithium, Li, has a positive one charge so one + sign can be placed in the element box. Beryllium, Be, has a positive two charge so ++ (two plus signs) can be placed in the Be box.

Name ______________________________ Date ____________

Activity 4: **The Metals Crossword**

Directions: Use the chemical symbols given in the "Across" and "Down" clues to determine each element name. Write the element name on the line by the chemical symbol for each metal. Then write the element name in the puzzle.

ACROSS

1. ____________________ (Co)
3. ____________________ (Ag)
6. ____________________ (Ti)
7. ____________________ (Na)
9. ____________________ (Ni)
13. ____________________ (Mg)
15. ____________________ (Fe)
16. ____________________ (Zn)
17. ____________________ (Ca)
19. ____________________ (Pb)

DOWN

2. ____________________ (Al)
4. ____________________ (Pt)
5. ____________________ (K)
8. ____________________ (Sn)
10. ____________________ (Cr)
11. ____________________ (Li)
12. ____________________ (Hg)
14. ____________________ (Au)
17. ____________________ (Cu)
18. ____________________ (Ir)

Name ________________________________ Date ____________

Activity 5: **Nonmetals Word Find**

Directions: Write the names and symbols of the seventeen nonmetals in the chart below. Then find each nonmetal name in the puzzle that follows and circle it. Answers can go up, down, or diagonally.

ELEMENT	SYMBOL	ELEMENT	SYMBOL	ELEMENT	SYMBOL

N	I	T	R	O	G	A	S	U	I	L	E	H	A	N
O	X	Y	A	A	E	R	U	B	R	A	C	Y	X	E
G	E	N	D	I	N	N	L	Z	Y	H	E	D	V	G
R	B	O	R	N	O	F	F	U	L	N	R	R	I	Y
A	P	N	A	N	E	X	U	O	I	A	T	O	O	X
K	H	Y	E	F	L	U	R	D	C	E	Z	G	D	O
R	O	X	I	O	E	I	O	A	N	V	O	E	I	B
Y	S	O	N	X	N	I	T	R	O	G	E	N	Z	K
M	U	I	N	E	L	E	S	O	B	R	O	O	M	R
T	L	L	C	H	L	O	N	U	L	F	H	O	X	Y
R	F	A	R	U	S	U	R	O	H	P	S	O	H	P
A	F	L	U	O	R	I	N	E	B	K	R	Y	O	T
D	H	Y	D	R	G	E	N	H	E	R	U	X	G	O
O	S	H	E	L	I	U	M	H	P	S	A	O	R	N
N	E	O	C	Y	B	R	O	M	I	N	E	C	A	R

Name ______________________________ Date ____________

Activity 6: **Mixed-Up Table**

Directions: Below is a section of the periodic table that is out of order. First, use the clue in each box to figure out the name of the element and write its name in the box. Then, find the section represented by the boxes below on the real periodic table. Write the element names in their proper order in the section marked "Corrected Table." Don't leave out any elements!

Aluminum 26.982	5	Ga	14.007	28.086
As	Br	32	15.999	32.066
Sb	12.011	49	30.974	52
(210)	Cl	53	84	81
83	18.998	Pb	Se	Sn

Corrected Table

Find **antimony, arsenic, astatine, boron, germanium, silicon,** and **tellurium** on your corrected chart. Underline each name with a green pen or pencil. They form an important pattern on the periodic table. These elements are the ______________________ .

Solid, Liquid, or Gas

Preliminary Handout:

- Solid, Liquid, or Gas (reproducible reading)

Activities:

7. Phases and Phase Change (hands-on demonstration)
8. Phases Puzzle (reproducible)

Solid, Liquid, or Gas

Some versions of the periodic table code the elements to indicate phase at room temperature.

- The elements that are ***liquids*** at room temperature are **bromine** and **mercury**. **Gallium** and **cesium** have melting points below 30°C or 86°F. Gallium and cesium are solids at room temperataure, but liquefy by the time the temperature reaches 30°C.
- The elements that are ***gases*** at room temperature are **hydrogen, helium, nitrogen, oxygen, fluorine, neon, chlorine, argon, krypton, xenon**, and **radon**.
- The **rest** of the elements are ***solids*** at room temperature.

Phase changes (changing from solid to liquid to gas) are physical changes. The element is the same element, just in a different form. The elements have distinct melting and boiling points; these physical characteristics are used to identify each element.

Activity 7: **Phases and Phase Change** (Hands-on Demonstration)

Objective: Students should be able to observe phase changes and recognize the phase of a given element.

Materials:

- Examples of different elements representing different phases, such as mercury, copper, iron, carbon, oxygen, helium
- Pan of water, food coloring, 1 meter of tubing, and a gas-collecting bottle
- Hotplate and tongs
- Glass beaker and watch glass
- Aluminum foil
- Sulfur, liquid nitrogen, ice, dry ice
- Overhead projector

Preparation: Have the different elements numbered and labeled with their element symbols.

Procedure:

1. Place the solid sulfur on a piece of aluminum foil. Then place the foil on the hotplate. Sulfur, when heated, will melt, forming liquid sulfur at 113°C.
2. Use the food coloring to color the water. Put colored water in the gas-collecting bottle. Place one end of the tubing in the gas-collecting bottle; then turn the bottle over into the pan of colored water. Place the other end of the tubing over the vial of liquid nitrogen. Open the vial and collect the nitrogen gas in the bottle. Liquid nitrogen, when exposed to room temperature, will form nitrogen gas. As nitrogen gas collects in the bottle, the water level in the bottle will decrease.
3. Although dry ice (solid carbon dioxide) and ice (solid water) are compounds and not elements, they can be used to show phase changes.
   (a) Students have all seen ice melt into water and water boil to form steam. Water neatly illustrates all three phases. Illustrate this by placing ice in the beaker on the hotplate. The ice will eventually melt and then boil. If the steam is captured and cooled, it will revert back to liquid water. If you want to capture steam, place ice in an Erlenmeyer flask fitted with a one-hole rubber stopper that has a short piece of glass tubing in the hole and 1-meter piece of rubber tubing attached. Place the rubber tubing into a gas-collecting bottle to collect the steam.
   (b) Not all substances go from solid to liquid to gas. Solid carbon dioxide undergoes sublimation and skips the liquid phase. Place a piece of dry ice on the watch glass on the overhead projector. Students will see the ice move around and see "smoke" rising. Dry ice sublimes into gaseous carbon dioxide without forming the liquid phase.
4. Have students write down the element symbol and the phase of the element for each of the samples.

Results: Students should be able to identify the phase of matter of different element samples and explain phase change.

Name ______________________ Date __________

Activity 8: **Phases Puzzle**

Directions: List the name and symbol of each element that is either a liquid or a gas at 30°C in the table below. In the puzzle, color the symbols for the liquid elements blue and the symbols for the gaseous elements red. What letters are formed by the symbols?

Liquids		***Gases***			
ELEMENT	SYMBOL	ELEMENT	SYMBOL	ELEMENT	SYMBOL

Tl	At	Bi	Se	Sn	Te	Ac	
Yb	Sr	Zn	Mn	Al	Li	Ra	Ce
Fm	Hg	Mg	Fr	S	B	Ti	Er
Nd	Br	C	Fe	Ca	K	V	Md
No	Cs	Ga	Hf	Si	Ge	Ta	Tm
Es	I	Cr	Mt	Ba	Rb	Zn	Pr
Cf	Cu	Sc	O	He	Na	Lr	U
Bk	Rf	Cl	W	Co	Ni	Tc	Cm
Pm	Rh	Ne	Pt	Be	Ag	Db	Tb
Pu	Pd	F	Ir	Ar	H	Nb	Am
Dy	Hs	Kr	Cd	Au	N	Os	Ho
Th	Re	Zr	Rn	Xe	Y	Lu	Np
Pa	Mo	Pb	Po	As	P	Ru	Gd
Sm	Bh	Sb	In	La	Eu	Sg	

Periodic Table: Patterns

Preliminary Handout:

- Periodic Table: Patterns (reproducible reading)

Activities:

9a. Mendeleev Song (teacher guide page)
9b. Mendeleev Quiz (reproducible)
10. Arrangement of Groups and Periods: Seeing Patterns (hands-on demonstration)
11a. Arrangement of Groups and Periods: Finding Periodicity Patterns (reproducible)
11b. Arrangement of Groups and Periods: Finding Number and Shape Patterns (reproducible)

Periodic Table: Patterns

Three major contributors to the establishment and current format of the periodic table of elements are Dmitri Mendeleev, Henry Moseley, and Glenn Seaborg. Dmitri Mendeleev (1834-1907), a Russian chemist, first published the periodic table of elements in 1869. Mendeleev's table was based on atomic mass, and although he could predict many of the elements that had not been discovered at that time, there were some minor problems with his version. Henry Moseley (1887-1915), an English physicist, took Mendeleev's table and rearranged the elements based on atomic number. Moseley's basic format is still used. Glenn Seaborg (1912-1999), an American nuclear chemist, changed the format by placing the lanthanide and actinide series below the table, which made it easier for publishers to print. This "actinide concept" was helpful in predicting properties of transuranium elements and played an important role in the discovery of these elements. Transuranium elements are elements with an atomic number greater than uranium's. Both Mendeleev and Moseley looked at the chemical and physical properties of the elements to group the known elements and to predict the existence of others. Seaborg, whose forte was transuranium elements, was one of the scientists who helped to make and study these synthetic elements.

Activity 9a: **Mendeleev Song**

Objective: Students will be able to recall historical information about Mendeleev and the original periodic law.

Materials:

- Recording of "Mendeleev Song" (You can find a recording of this song on *Chemistry Songbag*. Copies of the tape are available via email [scimusic@tranquility.net] or U.S. mail [Jeff Moran, c/o The Master Word Works, 2071 County Road 246, Fulton, MO 65251])
- Copy of quiz for each student

Procedure:

1. Have students listen to "Mendeleev Song" and ask them to list the information they recall about Mendeleev and the periodic law.
2. Play "Mendeleev Song" again and have students correct any incorrect information on their lists.
3. Hand out quiz.
4. Play "Mendeleev Song" for the third time and have students complete the quiz.

Results: Students should be able to correctly complete the quiz.

Name ______________________________ Date ____________

Activity 9b: **Mendeleev Quiz**

Directions: Write the answers to the questions below in the answer blanks.

1. Mendeleev's first name was ________________.
2. He proposed the periodic table of elements in the year ________________.
3. He based his periodic table on the element's atomic ________________.
4. He was from the country of ________________.
5. Element number ________________, Mendelevium, was named after Mendeleev.

Activity 10: **Arrangement of Groups and Periods: Seeing Patterns**

Objective: Each group of students will arrange the cards they are given into a pattern and predict the patterns seen on the ninth card.

Materials:
- Envelopes (1 per group)
- 3 x 5 index cards (9 per group) marked as follows:

Procedure: From each set remove one card, which the teacher keeps until the students have correctly described it. Place the remaining cards in an envelope. (Each envelope will have eight of the nine cards.)

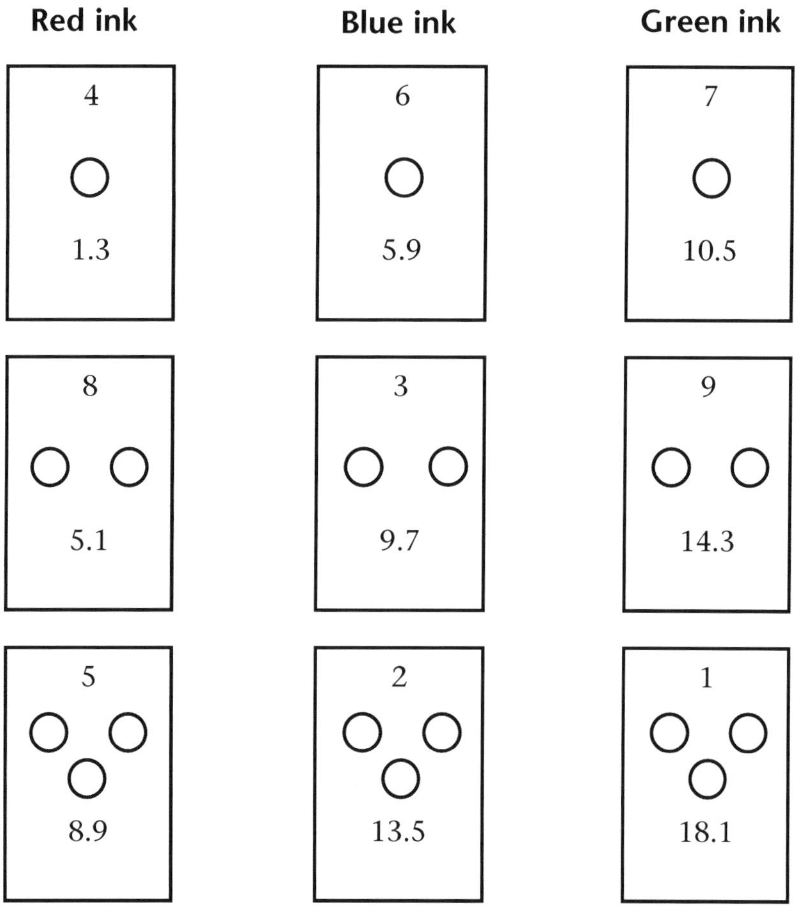

Results: Students should be able to describe the missing card by placing the cards from the envelope in the correct pattern (shown above). They should recognize the patterns to correctly place the cards into families and periods. The families are grouped by color and the bottom number increases from top to bottom by 3.8. The periods are grouped by the number of designs (circles) and the bottom number increases from left to right by 4.6. The top numbers do not form a pattern. Students are given the ninth card upon correctly describing it.

Name ______________________________ Date ____________

Activity 11a: **Arrangement of Groups and Periods: Finding Periodicity Patterns**

Directions: Read the following two problems to determine whether there are patterns (periodicity), and if there is periodicity, write down a probable cause for the periodicity.

Problem 1: Scientists studied the number of gazelles in the various game reserves in Zambia from 1979 until 1998. The population for each year was as follows:

YEAR	POPULATION	YEAR	POPULATION
1979	103	1989	674
1980	215	1990	850
1981	433	1991	1197
1982	695	1992	83
1983	883	1993	136
1984	1236	1994	265
1985	45	1995	512
1986	127	1996	692
1987	245	1997	897
1988	478	1998	1284

Problem 2: You live near an appliance factory and can see the trucks leaving the factory loaded with appliances for delivery to the distributors. You notice that more trucks leave at the end of March, June, September, and December than at any other time. In fact the number of trucks is especially high near the end of December.

Name ______________________________ Date ____________

Activity 11b: **Arrangement of Groups and Periods: Finding Number and Shape Patterns**

Directions: Look at the following number and shape patterns. Determine the pattern and predict the next number or shape.

1. 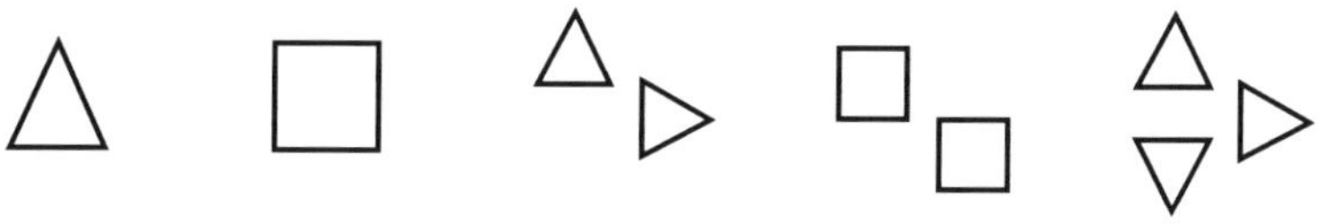 ____________________

2. 33, 11, 15, 5, 9 ____________________

3. 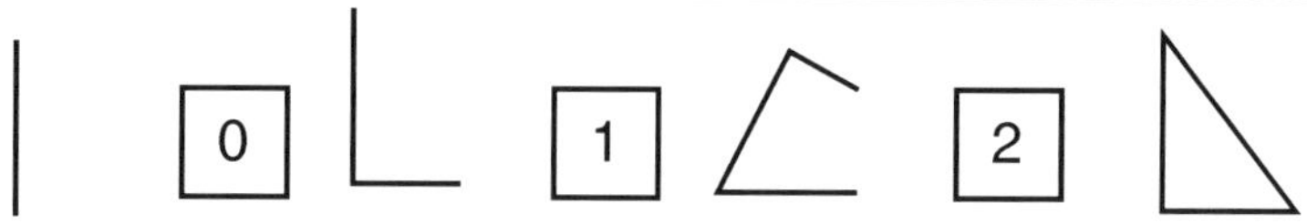 ____________________

4. 38, 37, 39, 36, 40, 35, 41, 34, 42 ____________________

5.  ____________________

6. 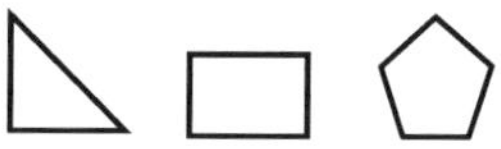 ____________________

7. 2, 4, 3, 9, 5, 25, 18, 324 ____________________

8. 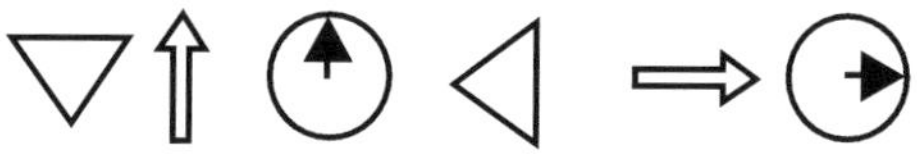 ____________________

9. 95, 87, 83, 89, 81, 77, 83, 75, 71, 77, 69, 65 ____________________

10. 11, 15, 31, 26, 1, 7, 43, 36, -13, -5 ____________________

Element Symbols and Other Stated Information

Preliminary Handout:

- Element Symbols and Other Stated Information (reproducible reading)

Activities:

12. An Elemental Trip Through Europe (reproducible)
13. Elemental Math (reproducible)
14. Cooking with the Elements (reproducible)
15. Matching Names and Symbols (reproducible)
16. Unusual Element Symbols (reproducible)

Element Symbols and Other Stated Information

Different versions of the periodic table of elements contain different information, but all versions contain the **element symbol** and the **atomic number**.

The **element symbol** consists of one or two letters. The first letter is always capitalized and the second letter is always lowercase. For elements that have been identified but not named, the symbol is three letters. These are derived from Greek and Latin words for the digits in the atomic number followed by *-ium.* The first is always a capital letter, and the other two letters are lower case letters. Therefore, element number 114 is ununquadium or Uuq.

Digit	Greek or Latin term
0	*nil*
1	*un*
2	*bi*
3	*tri*
4	*quad*
5	*pent*
6	*hex*
7	*sept*
8	*oct*
9	*enn*

The symbols represent the various element names and can be based on Latin names or the original foreign language name. There is one element, tungsten, whose symbol, W, is based on *wolfram* from the original German. There are ten elements whose symbols are based on their Latin names. The ten elements, their symbols and their Latin names are:

Element	Symbol	Latin Name
Antimony	Sb	*stibium*
Copper	Cu	*cuprum*
Gold	Au	*aurum*
Iron	Fe	*ferrum*
Lead	Pb	*plumbum*
Mercury	Hg	*hydrargyrum*
Potassium	K	*kalium*
Silver	Ag	*argentum*
Sodium	Na	*natrium*
Tin	Sn	*stannum*

The **atomic number** is the number of protons in the nucleus and is the identifying characteristic of the element.

Other types of information that may be indicated on the periodic table of elements are:

- element name
- mass number
- atomic mass
- electron configuration
- oxidation states
- phase at room temperature

There are versions of the periodic table of elements that indicate much more information—such as melting point, boiling point, density, atomic radius, and electronegativity. These more specific versions can be purchased from various scientific and chemical supply firms.

Name ______________________________ Date __________

Activity 12: **An Elemental Trip Through Europe**

Directions: For each element combination in parentheses below, the **symbols** for the elements form a word. Write the symbols in the answer blank following each group of elements. This will help you complete the story.

> **Example:** (*uranium + selenium*) = USe (or, the word "use").

(thorium + iodine + sulfur) __________ past summer we took a long *(vanadium + actinium + astatine + iodine + oxygen + nitrogen)* __________ through Europe. We saw many *(phosphorus + aluminum + actinium + einsteinium)* __________ during our trip. *(silicon + nitrogen + cerium)* __________ Europe is *(tungsten + helium + rhenium)* __________ much of *(americium + erbium + iodine + calcium + sulfur)* __________ cultural *(hydrogen + erbium + iodine + tantalum + germanium)* __________ is derived, we tried to see as many historical *(phosphorus + lanthanum + cerium + sulfur)* __________ as possible.

We spent *(nickel + neon)* __________ days in *(fluorine + radium + nitrogen + cerium)* __________ . In *(phosphorus + argon + iodine + sulfur)* __________ , after visiting the Eiffel Tower and the *(argon + carbon)* __________ de Triomphe, we *(sulfur + astatine)* __________ at a *(calcium + iron)* __________ to watch the people. We saw the cathedral at Chartres on our drive to the *(sulfur + oxygen + uranium + thorium)* __________ of *(fluorine + radium + nitrogen + cerium)* __________ . We took a short side trip *(oxygen + vanadium + erbium)* __________ the Pyrenees to Spain to visit Barcelona and *(oxygen + boron + tantalum + indium)* __________ a flavor of Castilian and Moorish influences. Then *(barium + carbon + potassium)* __________ to *(sulfur + oxygen + uranium + thorium + erbium + nitrogen)* __________ *(fluorine + radium + nitrogen + cerium)* __________ and the beautiful *(beryllium + actinium + helium + sulfur)* __________ of *(nickel + cerium)* __________ .

After seeing the French Riviera, we drove to Italy and through the *(tungsten + iodine + neon)* __________ country to Pisa. From Pisa it was a short drive following the *(argon + nobelium)* __________ River to Florence and the *(boron + iridium + thorium + phosphorus + lanthanum + cerium)* __________ of Renaissance art. Our next stop was the Eternal City, Rome. We had to see the Coliseum, the *(vanadium + astatine + iodine + calcium + nitrogen)* __________ and the Pope. From Rome we drove *(sulfur + oxygen + uranium + thorium)* __________ along the Italian coastline to *(sodium + polonium + lithium)* __________ and Pompeii. Then we headed northeast *(sulfur + indium + cerium)* __________ we could not leave Italy without a gondola ride on the canals of Venice.

From sea level at Venice, we headed for *(vanadium + erbium + yttrium)* __________ high ground as we drove *(oxygen + vanadium + erbium)* __________ the *(aluminum + phosphorus + sulfur)* __________ to Switzerland. We had *(holmium + phosphorus + einsteinium)* __________ of skiing down the *(aluminum + phosphorus + iodine + neon)* __________ slopes and *(aluminum + sulfur + oxygen)* __________ of buying a *(fluorine + americium + oxygen +*

(continued)

Activity 12: **An Elemental Trip Through Europe** *(continued)*

uranium + sulfur) __________ *(sulfur + tungsten + iodine + sulfur + sulfur)* __________ *(tungsten + astatine + carbon + hydrogen)* __________ . However, there was not enough *(tin + oxygen + tungsten)* __________ for skiing, so we headed for Austria and the *(silicon + tellurium + sulfur)* __________ from the movie *The (sulfur + oxygen + uranium + neodymium)* __________ *of Music.*

From Vienna and Salzburg and beautiful waltzes, we went to Germany to see the *(calcium + sulfur + thallium + einsteinium)* __________ of the *(boron + lanthanum + carbon + potassium)* __________ Forest and *(helium + argon)* __________ polkas. We survived the autobahn and visited the Peace *(gallium + tellurium)* __________ in *(boron + erbium + lithium + nitrogen)* __________ .

From *(boron + erbium + lithium + nitrogen)* __________ we headed *(boron + actinium + potassium)* __________ toward *(fluorine + radium + nitrogen + cerium)* __________ to *(chromium + osmium + sulfur)* __________ the English Channel. We decided to *(chromium + oxygen + sulfur + sulfur)* __________ by ferry so we could see the *(tungsten + hydrogen + iodine + tellurium)* __________ *(chlorine + iodine + fluorine + fluorine + sulfur)* __________ of Dover. We had *(nickel + neon)* __________ days left to see *(arsenic)* __________ much *(oxygen + fluorine)* __________ England as possible. We, of course, saw Buckingham *(protactinium + lanthanum + cerium)* __________ and the guards *(tungsten + iodine + thorium)* __________ their tall fuzzy *(boron + lanthanum + carbon + potassium)* __________ hats and stern *(fluorine + actinium + einsteinium)* __________ .

We caught a *(neon + tungsten)* __________ *(phosphorus + lanthanum + yttrium)* __________ at Covent Gardens and a Shakespearean *(phosphorus + lanthanum + yttrium)* __________ at Stratford-on-Avon. We drove as *(fluorine + argon)* __________ north as Sherwood Forest and Nottingham to relive the *(fluorine + americium + oxygen + uranium + sulfur)* __________ tales of Robin Hood. Then *(sulfur + oxygen + uranium + thorium)* __________ and west to see Stonehenge on the Salisbury *(phosphorus + lanthanum + indium + sulfur)* __________ and the *(beryllium + actinium + helium + sulfur)* __________ at Bournemouth and *(sulfur + oxygen + uranium + thorium + americium + platinum + oxygen + nitrogen)* __________ .

(boron + yttrium) __________ then it *(tungsten + arsenic)* __________ time to return to London's Heathrow Airport for the *(phosphorus + lanthanum + neon)* __________ flight *(barium + carbon + potassium)* __________ to *(americium + erbium + iodine + calcium)* __________ . *(tungsten + hydrogen + astatine)* __________ an amazing *(vanadium + actinium + astatine + iodine + oxygen + nitrogen)* __________ ! We will *(neon + vanadium + erbium)* __________ forget our elemental trip through Europe.

Name ______________________________ Date ____________

Activity 13: **Elemental Math**

Directions: Replace the symbol for each element below with the correct atomic number. Then complete the equation. Finally, translate the answer back into the element symbol.

> **Example:** H + He = _____ would be calculated as 1 + 2 = 3. 3 is the atomic number for lithium, so the equation would be H + He = Li.

1. Cl + He = ______________________
2. Tc + Ag – Ne = ______________________
3. (H + Br) ÷ Li = ______________________
4. (Cs + Pa) ÷ He = ______________________
5. Na × Be + C = ______________________
6. Re ÷ (Sc + Be) = ______________________
7. Fm ÷ Mn × Be = ______________________
8. In ÷ N × B + Es – Mn = ______________________
9. C^{He} × Ne ÷ Mg = ______________________
10. Te + Xe – Sg + As + Pd = ______________________
11. (Pm – Sb) × O + F = ______________________
12. [(Zr + Ge) ÷ F] + [Xe ÷ C × Li – Mg] = ______________________

Name ______________________________ Date __________

Activity 14: **Cooking with the Elements**

Directions: For each element combination in parentheses below, use the **symbols** for the elements to obtain a scrambled word. Then unscramble the letters to form the correct words. Write the symbols in the answer blank following each group of elements. This will help you complete each numbered paragraph.

> **Example:** *(boron, indium, oxygen, tantalum)* = BInOTa, which unscrambles to form the word OBTaIn.

1. For breakfast we *(yttrium + francium)* __________ eggs, *(cobalt + nitrogen + barium)* __________ and *(hydrogen + hydrogen + arsenic)* __________ *(oxygen + nitrogen + tungsten + bromine)* __________ potatoes, and toast *(astatine + tungsten + helium)* __________ or *(hydrogen + tellurium + tungsten + iodine)* __________ bread. Or, we can have *(nitrogen + calcium + einsteinium + protactinium + potassium)* __________ or waffles and sausage, or *(aluminum + cerium + rhenium)* __________ , such as *(radon + cobalt)* __________ *(lanthanum + potassium + fluorine + einsteinium)* __________ or *(nitrogen + iodine + silicon + radium)* __________ *(boron + nitrogen + radium)* __________ , with milk.

2. *(thorium + helium + aluminum + yttrium)* __________ *(potassium + actinium + sulfur + tin)* __________ would be fruits, such as *(sodium + sodium + barium + sulfur)* __________ , grapes, *(sulfur + iodine + tungsten + potassium + iodine)* __________ , apples, and oranges and different *(einsteinium + carbon + helium + einsteinium)* __________ and *(potassium + chromium + erbium + actinium + sulfur)* __________ . Of course, most of us would *(erbium + radium + thorium)* __________ have *(hydrogen + phosphorus + sulfur + carbon + iodine)* __________ , *(iodine + oxygen + cobalt + potassium + einsteinium)* __________ , or *(nitrogen + dysprosium + calcium)* __________ .

3. For drinks, we *(fluorine + phosphorus + rhenium + erbium)* __________ *(calcium + cobalt + lanthanum + cobalt)* __________ or another type of soda *(vanadium + erbium + oxygen)* __________ milk, juice or *(erbium + astatine + tungsten)* __________ .

4. Most people have fast food and *(selenium + uranium)* __________ the drive *(ruthenium + sulfur + thorium)* __________ for lunch. They usually have only half an hour and *(oxygen + carbon + selenium + holmium)* __________ *(carbon + tantalum + osmium)* __________ or hamburgers and French *(einsteinium + iodine + francium)* __________. Sometimes they will be *(carbon + yttrium + lutetium + potassium)* __________ and have a salad, *(uranium + phosphorus + sulfur + oxygen)* __________ , sandwich, or *(neon + iodine + hydrogen + selenium + carbon)* __________ take-out. At *(erbium + oxygen + thorium)* __________ times, people, especially students, eat *(holmium + sulfur + carbon + sodium)* __________ or *(carbon + lithium + iodine + hydrogen)* __________ cheese *(iodine + francium + einsteinium)* __________ .

(continued)

Name ______________________________ Date ____________

Activity 14: **Cooking with the Elements** *(continued)*

5. Dinners are the big meals. *(iodine + sulfur + thorium)* __________ is *(helium + tungsten + nitrogen)* __________ families *(thorium + gallium + erbium)* __________ together after a long day. Dinners usually consist of a main dish containing some type of meat. The meat can be *(neon + terbium + oxygen)* __________ , *(americium + hydrogen)* __________ , pork *(sulfur + phosphorus + carbon + holmium)* __________ , chicken, *(boron + barium + yttrium)* __________ *(carbon + barium + potassium)* __________ ribs, prime rib, or *(iodine + hydrogen + sulfur + fluorine)* __________ .

6. Of course, there is always some type of carbohydrate. *(iodine + thorium + sulfur)* __________ is usually a potato, which we can bake, mash, *(yttrium + francium)* __________ , scallop, or boil. For variety, there is also rice or *(tantalum + arsenic + phosphorus)* __________ .

7. There usually is a *(holmium + cerium + iodine + carbon)* __________ of vegetables. Some *(sulfur + carbon + iodine + holmium + cerium)* __________ are *(radon + cobalt)* __________ , peas, *(cobalt + lithium + oxygen + bromine + carbon)* __________ , beans, *(silver + arsenic + phosphorus + uranium + argon + sulfur)* __________ or squash.

8. One of my favorite *(sulfur + uranium + sulfur + oxygen + phosphorus)* __________ is *(tungsten + neon)* __________ England *(americium + chlorine)* __________ chowder. I *(gold + sulfur + tellurium)* __________ the *(nitrogen + barium + cobalt)* __________ and *(nitrogen + oxygen + nickel + oxygen + sulfur)* __________ first. Then I add *(astatine + tungsten + erbium)* __________ , *(sulfur + chlorine + americium)*__________ , celery, and *(iodine + sulfur + sulfur + phosphorus + cerium)* __________ such as *(yttrium + barium)* __________ leaf, thyme, and marjoram. The diced potatoes and *(rhenium + carbon + americium)* __________ are added about thirty minutes *(oxygen + rhenium + beryllium + fluorine)* __________ serving.

9. The best part is dessert. There are many different *(potassium + calcium + einsteinium)* __________ and *(iodine + einsteinium + phosphorus)* __________ . *(uranium + rubidium + rhodium + barium)* __________ looks like red celery, and is tart *(helium + tungsten + nitrogen)* __________ baked in *(phosphorus + einsteinium + iodine)* __________ . Another simple dessert is *(cerium + iodine)* __________ *(americium + carbon + rhenium)* __________ . *(iodine + thorium + sulfur)* __________ can be served with *(erbium + oxygen + thorium)* __________ desserts or *(yttrium + boron)* __________ itself, in a *(neon + cobalt)* __________ or a dish, *(iodine + thorium + tungsten)* __________ *(cobalt + tellurium + carbon + lanthanum + holmium)* __________ or *(yttrium + neodymium + calcium)* __________ toppings.

10. A fancy dessert is *(phosphorus + rhenium + carbon + einsteinium)* __________ with a *(neodymium + boron + radium + yttrium)* __________ sauce. Many *(neon + iodine + fluorine)* __________ restaurants will *(cerium + lanthanum)* __________ their desserts with *(nitrogen + boron + dysprosium + radium)* __________ to make a flaming dessert.

Name ______________________________ Date ____________

Activity 15: **Matching Names and Symbols**

Directions: Match each element symbol with its element name. Place the letter(s) of the correct symbol on the line in front of each numbered element.

ELEMENT NAMES		ELEMENT SYMBOLS
_____	1. Lithium	P
_____	2. Aluminum	Mg
_____	3. Neon	Cl
_____	4. Calcium	O
_____	5. Sulfur	Mn
_____	6. Boron	Al
_____	7. Chromium	Ne
_____	8. Zinc	N
_____	9. Helium	Ca
_____	10. Phosphorus	Zn
_____	11. Oxygen	B
_____	12. Chlorine	C
_____	13. Magnesium	Ar
_____	14. Manganese	He
_____	15. Hydrogen	H
_____	16. Iodine	Li
_____	17. Silicon	Cr
_____	18. Argon	S
_____	19. Carbon	Si
_____	20. Nitrogen	I

Name ______________________________________ Date ______________

Activity 16: **Unusual Element Symbols**

Eleven elements have names that are unlike their symbols. This is because ten of the symbols are based on Latin words, and one is based on a German name (see example). By decoding the numbers below, you can find the Latin names for these symbols.

Directions: Number the letters of the alphabet from 1 to 26. For example, A is 1, B is 2, etc. Then write the letter each number represents on the line above that number. After you are finished decoding the numbers, write the common element name next to its symbol.

	SYMBOL	ELEMENT	GERMAN NAME
Example:	W	<u>tungsten</u>	<u>W</u> <u>O</u> <u>L</u> <u>F</u> <u>R</u> <u>A</u> <u>M</u> 23 15 12 6 18 1 13

SYMBOL	ELEMENT	LATIN NAME
1. Ag	________	1 18 7 5 14 20 21 13
2. Au	________	1 21 18 21 13
3. Cu	________	3 21 16 18 21 13
4. Fe	________	6 5 18 18 21 13
5. Hg	________	8 25 4 18 1 18 7 25 18 21 13
6. K	________	11 1 12 9 21 13
7. Na	________	14 1 20 18 9 21 13
8. Pb	________	16 12 21 13 2 21 13
9. Sn	________	19 20 1 14 14 21 13
10. Sb	________	19 20 9 2 9 21 13

Protons, Neutrons, and Electrons

Preliminary Handout:

- Protons, Neutrons, and Electrons (reproducible reading)

Activities:

17. Paper Atomic Models (hands-on activity)
18. Three-Dimensional Models (hands-on activity)
19. Calculating Protons, Neutrons, and Electrons Given A and Z (reproducible)
20. Metal Ball Drop (hands-on activity)

Protons, Neutrons, and Electrons

Protons are positively charged particles with a mass of approximately 1 **atomic mass unit** (amu). They are found in the nucleus of an atom. Each element is identified by the number of protons in its nucleus.

All atoms with only **1** proton in their nuclei are **hydrogen atoms**. All atoms with 92 protons in their nuclei are **uranium atoms**. The number of protons is known as the **atomic number** and is represented by the symbol Z.

Another major particle, also found in the nucleus, is the **neutron**. Neutrons are neutral particles with a mass of approximately 1 amu (same as protons). The **atomic mass number**, represented by **A**, is the sum of the number of protons and neutrons.

The third major particle of the atom is the **electron**. Electrons are negatively charged particles. In an electrically neutral atom, the number of electrons equals the number of protons. An electron has a mass of approximately 1/1836 amu, or 5×10^{-4} amu. Electrons are found outside the nucleus in one of various energy levels.

Energy levels are regions within the atom to which electrons are restricted. They have also been called orbits and and shells to make them easier to understand. The location of the electrons in the various energy levels is responsible for the chemical properties of the elements.

Electrons in the different energy levels have different amounts of energy. If an electron gains energy, for example, it will "jump" to a higher energy level. As electrons drop from higher energy levels to their lowest energy level **(ground state)**, they emit differing amounts of energy. This energy is emitted as different wavelengths of the electromagnetic spectrum.

Activity 17: **Paper Atomic Models**

Objective: Students will construct models reflecting the atomic structure for various elements. The correct number of protons and neutrons should appear in the nucleus and the correct number of electrons should appear in each energy level.

Materials:
- large (20-cm diameter) paper circles to represent the atom
- small (3-cm diameter) paper circles to represent the nucleus
- circle punches in two different colors to represent protons and neutrons
- star punches to represent electrons

Procedure: Give each group at least three large and three small circles per group, as well as 100-150 of each of the three punches. Assign each group several different atoms to construct. The elements from hydrogen to krypton will provide students with enough examples.

Results: Students should be able to construct accurate Bohr models of various atoms. They should place the small circle in the middle of the large circle, the correct number of protons and neutrons in the small circle, and lastly, the correct number of stars (electrons) in the appropriate energy level (i.e., a maximum of 2 in the first energy level, a maximum of 8 in the second energy level and a maximum of 8 in the third energy level).

Activity 18: **Three-Dimensional Models**

Objective: Students will creatively construct models of various elements.

Materials: Students supply their own materials.

Procedure: Have students draw numbers to determine which element they will construct. They should construct a three-dimensional model of the element selected, using whatever materials they desire. The model should accurately reflect the nucleus with protons and neutrons and the electrons in various energy levels.

Results: This is another method of evaluating students' understanding of basic atomic structure without paper-and-pencil tests.

Extension: As a variation, have students explain how their models reflect isotopes and ions.

Name ______________________________ Date ____________

Activity 19: **Calculating Protons, Neutrons, and Electrons Given A and Z**

Directions: Complete the table. Assume that the atoms are neutral when calculating electrons.

ELEMENT	ELEMENT SYMBOL	ATOMIC NUMBER (Z)	MASS NUMBER (A)	NUMBER OF PROTONS	NUMBER OF NEUTRONS	NUMBER OF ELECTRONS
Carbon		6			6	
Silicon			28	14		
	Fe	26	56			
Gold				79	118	
	Ag				61	47
	Pb	82			125	
Fluorine			19	9		
Oxygen		8	16			
	Mg			12	12	
	K	19			20	
Copper			64			29
Nitrogen		7	14			
Hydrogen		1	1			
	Na			11	12	

Activity 20: **Metal Ball Drop**

Objective: Students "see" electron energy drops and their effects.

Materials:
- Metal balls
- Modeling clay
- Flat, shallow pans
- Blindfolds (safety goggles with paper taped over them)

Preparation:
1. Place modeling clay into flat, shallow pans.
2. Measure and mark vertical distances of 10 cm, 20 cm, and 30 cm from the top of the clay.

Procedure:
1. Ask the question: "How would you determine the height from which a ball is dropped if you cannot see it being dropped?"
2. Write down various responses.
3. Once students are blindfolded, drop metal balls from 10 cm, 20 cm, and 30 cm into a pan of modeling clay and label the holes as A, B, and C. (The letters should not directly correspond to the heights.)
4. Have students match the hole with the vertical distance.

Results: Students should be able to explain the relationship between the dropping metal balls and electrons dropping from higher energy levels to their ground state. The dropping ball has **kinetic energy** which it "gives up" when it strikes the clay. The released kinetic energy produces the hole in the clay. The ball with the most kinetic energy (i.e., falling the farthest distance) gives off the most energy and, therefore, makes the deepest hole.

Conclusions: Students can write answers to the following questions or the questions can be used to lead a class discussion.

1. Explain the differences in the three holes.
2. How can you determine the height from which the ball is dropped if you cannot see the ball dropped?
3. How does the dropped ball simulate electrons moving from the "excited" state to the ground state?

Isotopes and Ions

Preliminary Handout:

- Isotopes and Ions (reproducible reading)

Activities:

21a. Ion Charges (teacher guide page)
21b. Ion Charges (reproducible)
22. Isotopes and Ions (reproducible)
23. Combining Game (hands-on activity)

Isotopes and Ions

Although all atoms of a particular element have the same number of protons, all atoms do not have the same number of neutrons. Atoms of the same element with varying numbers of neutrons are known as **isotopes**. Isotopes are **neutral atoms**, which means the number of protons and electrons in the atom are equal. Isotopes of the same element differ according to the number of neutrons found in the nucleus. For example, hydrogen, the lightest element, has three isotopes. The most common of these isotopes has one proton and no neutrons. Of the other two isotopes, deuterium has one proton and one neutron, and tritium, which is radioactive, has one proton and two neutrons. All the other elements also have isotopes; however, these are not given special names, but are known by their mass number; such as carbon-14 or uranium-238.

Atoms are neutral or have no electrical charge since the number of protons (positive charges) equals the number of electrons (negative charges). An ion (an electrically charged atom) is formed when the atom either gains or loses electrons. If the atom gains an electron, a negative ion or **anion** is formed, because there are more electrons (negative charges) than protons (positive charges). If the atom loses an electron, a positive ion or **cation** is formed, because there are more protons (positive charges) than electrons (negative charges). The atom can gain or lose more than one electron depending on the element.

The periodic table of elements can be used to determine the charge on the ion.

- All elements in group 1 form ions that have a 1+ charge.
- All elements in group 2 form ions that have a 2+ charge.
- All elements in group 13 form ions that have a 3+ charge.
- All elements in group 15 form ions that have a 3– charge.
- All elements in group 16 form ions that have a 2– charge.
- All elements in group 17 form ions that have a 1– charge.
- All elements in group 18 do not normally form ions, and therefore have 0 charge.
- The elements of groups 3 through 12, the transition elements, form ions with a variety of charges.

Activity 21a: **Ion Charges**

Objective: Students will learn the ion charges associated with the different groups in the periodic table.

Materials:

- Copy of the periodic table for each student
- Koosh and a separate list of element names or a beach ball with element symbols on it. (If using a beach ball, write the element symbols for elements in groups 1, 2, 13, 15, 16, 17, and 18 randomly on the beach ball. If using a koosh, compile a random listing of the element names for the elements in groups 1, 2, 13, 15, 16, 17, and 18.)
- Copy of worksheet for each student

Procedure:

1. Have students write the appropriate ion charge in each element block for elements in groups 1, 2, 13, 15, 16, 17, and 18. Charges are as follows:

Group Number	Ion Charge
1	1+
2	2+
13	3+
15	3-
16	2-
17	1-
18	0

2. Toss koosh or beach ball to a student. If using the koosh, the student catching the koosh must state the ion charge for the first element on the previously compiled list. The next student will identify the ion charge for the second element on the list, and so on. If using the beach ball, the student catching the ball must state the ion charge for the element whose symbol is under his or her left thumb. Continue until all students have at least one chance and all elements in the seven groups have been covered.
3. Have students complete the worksheet on the next page.

Results: Students should be able to recognize the correlation between the group number and the ion charge of the elements.

Name ______________________________ Date ____________

Activity 21b: **Ion Charges**

Directions: Write the ion charges for the following elements.

ELEMENT	ION CHARGE
Fluorine	
Magnesium	
Oxygen	
Phosphorus	
Hydrogen	
Neon	
Chlorine	
Aluminum	
Sodium	
Calcium	
Lithium	
Bromine	
Helium	
Nitrogen	
Rubidium	
Barium	
Potassium	
Boron	
Sulfur	
Iodine	

Name ______________________________ Date ____________

Activity 22: **Isotopes and Ions**

Directions: Use the data given to complete the table. Isotopes are also neutral atoms. All atoms of the same element (e.g., carbon-14 and carbon-12) are isotopes of that element. Consider isotopes with mass numbers equal to the mass numbers stated on the periodioc table of elements as the neutral atom. Therefore, the isotopes with mass numbers that are different from the mass numbers stated on the periodic table of elements will be considered isotopes for this activity. For ions, indicate the positive or negative charge.

ELEMENT	SYMBOL	ATOMIC NUMBER (Z)	MASS NUMBER (A)	NUMBER OF PROTONS	NUMBER OF NEUTRONS	NUMBER OF ELECTRONS	NEUTRAL ISOTOPE, POSITIVE ION NEGATIVE ION
Carbon		6	12				Neutral
				6	8	6	
Oxygen		8			8	10	
	Mg	12	24			10	
Uranium		92	238				Neutral
	U				143	92	Isotope
Helium		2	4			2	
	Na	11			12	10	
Chromium		24	52				2+ ion
	Cl	17			18		1- ion
	Hg	80	201				Neutral
Aluminum				13	14		3+ ion
	Ne		20	10			Neutral
		1	2				Isotope
				1	0		1+ ion
Phosphorus				15	16	15	
	Li				4	3	Neutral
Potassium				19	20	19	
	Fe	26			30	23	

Activity 23: **Combining Game**

Objective: Students will be able to visualize how ion charges determine the way elements combine to form new compounds.

Materials:

- Several "decks" of 3x5 index cards with element symbols and charges, for example, Cu^{2+}, Cu^{+}, O^{2-}, N^{3-}, etc. (Make sure there are enough duplicates to allow the same compounds to be formed more than once.)
- Several different lists of compounds to be formed, for example, CuS, Cu_2S, Al_2O_3, etc.

Procedure:

1. Divide students into groups of 3 to 5.
2. Give each group one "deck" of index cards and one list of compounds.
3. Have one student deal 7 cards to each member of the group. The rest of the "deck" serves as the source of new cards.
4. Students should use the cards in their hands to make as many of the compounds listed as possible. For example, if the compound is Li_2O, the student must place two Li+ cards and one O^{2}- card together.
5. Once students have formed their initial compounds the game basically follows the rules of the card game "Go Fish." The play starts with the dealer asking the student on his/her right for a specific card by stating the element name and charge. If that student has the requested card, he/she must relinquish the card. If the student on the right does not have the requested card, the student requesting the card draws a card from the "deck."
6. Play continues until a student runs out of cards. The winner is the student who has the most compounds.

Results:

- This reinforces element symbols, ion charges, and subscripts.
- Students also learn several binary compounds.

Extensions:

1. Instead of listing formulas on the compound list, use the compound names. This will give students practice in determining formulas.
2. Have students write down the formulas and names of the compounds they make.
3. Have students make up their own lists of compounds.

Periodicity

Preliminary Handout:

- Periodicity (reproducible reading)

Activities:

24. Periodicity in Groups—Magnet Demonstration (hands-on demonstration)
25. Periodicity in Periods—Magnet Demonstration (hands-on demonstration)
26. Periodic Properties (hands-on activity)
27. Periodic Riddles (reproducible)

28a. Periodic Table of Extraterrestrial Elements (teacher guide page)

28b. Periodic Table of Extraterrestrial Elements (reproducible)

Periodicity

By knowing how to use the periodic table of elements, we can obtain considerable information about elements. This information is based on periodicity and the format of the periodic table.

The vertical columns of the periodic table represent elements with similar chemical properties. These are called **groups** or **families**. There are several distinct groups or families of elements. Some of them are the alkali metals, alkaline earth metals, nitrogen family, oxygen family, halogens, and noble gases.

Within each group or family, the **atomic radii** increase from top to bottom of the periodic table. The atomic radius is the size of the atom from the center of the nucleus to the outermost principal electron energy level or valence shell. Reading down a group, each horizontal row, or **period**, adds an electron energy level. Because there are more electron energy levels, the size of the atom or atomic radius increases.

Two other properties of atoms decrease from top to bottom within a group on the periodic table. One of these properties is **ionization energy**. Ionization energy is the energy needed to remove the most loosely held electrons. These are the electrons that are the farthest from the nucleus. Ionization energy measures how easily an electron can be taken from an atom. The nucleus containing the protons is the positive charge of the atom. Each electron is a negative charge. The larger the atom, the greater the distance to the outermost electron energy level. Since the distance from the nucleus to the outermost electrons increases as the atomic radius increases, the ionization energy decreases. Although there are more protons (more positive charges) in the nucleus, the distance between the positively charged nucleus and the negatively charged electrons results in a decrease in the attractive power. The farther the electron is from the nucleus, the easier it becomes to remove it from the atom.

Electronegativity will also decrease as you read down the periodic table within a single group. Electronegativity is the ability of the atom to attract electrons. Electronegativity indicates how well an atom is able to take an electron from another atom. The electrons in the outermost or highest principal energy level are known as the **valence electrons**. The valence electrons are found in the *s* and *p* **sublevels**, or divisions, of the highest principal energy level. The more **kernel electrons** (electrons other than the valence electrons) between the nucleus and the valence electrons, the greater the negative charge. As the negative charge increases, so does the repulsion between the kernel and valence electrons. This repulsion is kown as a **shielding effect**. As the shielding effect increases, there is less positive attraction to pull electrons from another atom.

Going from left to right across a period, the atomic radii decrease, while the ionization energy and electronegativity increase. Each period represents a principal electron energy level. Although the number of electrons increases across a period, there is a greater attraction pulling the electrons toward the nucleus, resulting in a decrease in the atomic radius. Since the atomic radius decreases, the outermost electrons are held more strongly. This leads to an increase in ionization energy. It is harder to take an electron away from the atom. A greater positive attraction also increases the ability of the nucleus to take additional electrons. In other words, it increases the electronegativity.

Activity 24: **Periodicity in Groups—Magnet Demonstration**

Objective: Demonstrate that greater distance decreases the ability of the nucleus to hold and attract electrons.

Materials:

- Several magnets to represent the atomic nuclei (magnets should be same size and same strength, if possible)
- Several small nails or tacks to represent electrons
- Overhead projector and transparency with five concentric circles

Preparation:

1. Test magnets with tacks to determine the radii necessary to assure a successful demonstration.
   a. Place one magnet in the center of a blank transparency. Moving a tack toward the magnet, indicate the point where the magnet attracts the tack. Repeat several times to assure that you have determined the correct radius.
   b. Place two magnets side-by-side in the center of the circles. Moving a tack toward the magnets, indicate the point where the magnets attract the tack. Repeat several times to determine the correct radius.
   c. Continue the same procedure with three and four magnets.
2. Indicate the center spot for the nucleus, and draw five concentric circles to represent the electron energy levels on the transparency using radii obtained through step 1. This becomes the transparency with five concentric circles listed under **Materials.**

Procedure: To demonstrate using the alkali metals as an example, set up the following on the overhead. As additional magnets are added to the center (nucleus), they should be placed next to each other (side-by-side).

1. For lithium, place two tacks on the innermost circle and one tack on the second circle. Place one magnet in the center. The two tacks on the innermost circle should be attracted to the magnet. The third tack may move slightly, but should not be attached to the magnet.
2. For sodium, place two tacks on the innermost circle, eight tacks on the second circle, and one tack on the third circle. Place two magnets in the center. The tacks on the first two circles should be attracted to the magnet. The tack on the third circle should not be attached to the magnet.

(continued)

3. For potassium, place two tacks on the innermost circle, eight tacks on both the second and third circles and one tack on the fourth circle. Place three magnets in the center. The tacks on the first three circles should be attracted to the magnet. The tack on the fourth circle should not be attached to the magnet.
4. For rubidium, place two tacks on the innermost circle, eight tacks on both the second and third circles, eighteen tacks on the fourth circle and one tack on the fifth circle. Place four magnets in the center. The tacks on the first four circles should be attracted to the magnet. The tack on the fifth circle should not be attached to the magnet.

Results: Students should see how the greater distance from the nucleus allows the outermost electrons to be easily removed even though there are more protons (positive charges) in the nucleus attempting to hold the electrons. The farther away the electrons are, the weaker the attraction.

Extension: The shielding effect of electrons can be demonstrated by having students remove one tack from a magnet that has three tacks on it. Then compare the amount of force needed to remove the outermost tack from a magnet that has 38 tacks on it.

Activity 25: **Periodicity in Periods— Magnet Demonstration**

Objective: Demonstrate how the number of protons in the nucleus affects the nucleus's ability to hold and attract electrons.

Materials:

- Several magnets to represent the atomic nuclei (magnets should be same size and same strength, if possible)
- Several small nails or tacks to represent electrons
- Overhead projector and transparency with circle indicating atomic radius

Preparation:

1. Test magnets with tacks to determine the radius necessary to assure a successful demonstration.
   a. Place one magnet in the center of a blank transparency. Moving a tack toward the magnet, indicate the point where the magnet attracts the tack. Repeat several times to assure that you have determined the correct radius.
   b. Place two magnets side-by-side in the center of the transparency. Verify that two tacks placed on the radius will move toward the magnets. Repeat several times to assure that the tacks are slightly attracted to the magnets. This will indicate a smaller atomic radius.
   c. Continue the same procedure, adding one magnet at a time, until you use eight magnets. Tacks should not be completely attracted to the magnets until there are eight magnets in the center. If the tacks are completely attracted before the addition of the eighth magnet, restart the activity with a larger radius.
2. On the transparency, indicate the center spot for the nucleus. Draw a circle to indicate the atomic radius, using the radius obtained through step 1. This becomes the transparency listed under **Materials**.

Procedure: To demonstrate using period 2 as an example and showing only valence electrons, set up the following on the overhead. As additional magnets are added to the center (nucleus), they should be placed next to each other (side-by-side).

1. For lithium, place one tack on atomic radius and one magnet in the center. The tack should not move.
2. For beryllium, place two tacks on the atomic radius and two magnets in the center. The tacks should move slightly toward the magnets. This indicates a slightly smaller atomic radius that can be measured or marked.
3. For boron, place three tacks on the original atomic radius and three magnets in the center. The tacks should move toward the magnets. This atomic radius should be smaller than beryllium's.

(continued)

4. For carbon, place four tacks on the original atomic radius and four magnets in the center. The tacks should move toward the magnets. This atomic radius should be smaller than boron's.
5. Continue through neon, measuring or marking the atomic radius of each element.

Results: The students should see that the addition of protons (more positive attraction) to the nucleus decreases the atomic radii, because the electrons are pulled closer.

Activity 26: **Periodic Properties**

Objective: Students will recognize the properties that tend to illustrate periodicity in both periods and groups.

Materials:
- 22 strips of poster board (10 cm × 70 cm)
- 400 green toothpicks to represent atomic radius
- 400 red toothpicks to represent Pauling electronegativity
- 400 blue toothpicks to represent first ionization energy
- Reference books listing Pauling electronegativity, first ionization energy and atomic radii for the elements

Preparation:

1. On six strips of poster board, mark seven 10 cm × 10 cm squares. On the back of the six strips, identify as follows:
   - First strip — 1
   - Second strip — 18
   - Third strip — L1
   - Fourth strip — L2
   - Fifth strip — A1
   - Sixth strip — A2
2. On six strips of poster board, mark six 10 cm × 10 cm squares, starting 10 cm from left edge. On the back of the six strips, identify as follows:
   - First strip — 2
   - Second strip — 13
   - Third strip — 14
   - Fourth strip — 15
   - Fifth strip — 16
   - Sixth strip — 17
4. On ten strips of poster board, mark four 10 cm × 10 cm squares, starting 30 cm from left edge. On the back of the ten strips, identify as follows:
   - First strip — 3
   - Second strip — 4
   - Third strip — 5

(continued)

- Fourth strip — 6
- Fifth strip — 7
- Sixth strip — 8
- Seventh strip — 9
- Eighth strip — 10
- Ninth strip — 11
- Tenth strip — 12

Procedure:

1. Divide students into pairs. Have each pair take a strip of poster board.
2. The numbers on the back of the strips represent numbers of the groups (or families) on the periodic table. In addition, L1 will be the first seven elements of the lanthanide series (cerium through gadolinium). L2 will be the last seven elements of the lanthanide series (terbium through lutetium). A1 will be the first seven elements of the actinide series (thorium through curium) and A2 will be the last seven elements of the actinide series (berkelium through lawrencium).
3. Students should complete a square for each element in the family by filling in the following information:
    - Element symbol
    - Atomic number
    - Atomic radius
    - Electronegativity
    - First ionization energy
4. Students should cut the green toothpicks into the appropriate length to depict the relative atomic radii of each element. A whole toothpick will represent an atomic radius of 270. Approximately .25 toothpick will represent the smallest atomic radius of 70. Then, they should glue the green toothpicks next to the numerical value of the atomic radii, so the length of the toothpick is perpendicular to the poster board.
5. Students should cut the red toothpicks into the appropriate length to depict the electronegativity of each element. A whole toothpick will represent 3.98 and approximately .17 toothpick will represent the smallest electronegativity of 0.7. Then, they should glue the red toothpicks next to the numerical value of the electronegativity, so the length of the toothpick is perpendicular to the poster board.

(continued)

6. Students should cut the blue toothpicks into the appropriate length to depict the first ionization energy of each element. A whole toothpick will represent 2372.3 kJ/mol and approximately .17 toothpick will represent the smallest first ionization energy of 400 kJ/mol. Then, they should glue the blue toothpicks next to the numerical value of the first ionization energy so the length of the toothpick is perpendicular to the poster board.
7. Once the students have completed their strips, the total periodic table can be put together on the bulletin board.

Results: Students should see the periodic trends in the groups on their individual strips. Once the strips are put together to form the periodic table, students should see the periodic trends in the periods (horizontal rows). Students with the lanthanide and actinide series will only see part of the period trend on their individual strips.

Extensions:

1. Have each group graph the trends for their elements on four separate transparency sheets. Use green for atomic radii, red for electronegativity, blue for first ionization energy, and black for atomic number. Overlay the transparencies to show trends within a group and to compare trends among the groups. Place element symbols on the x-axis and the numerical value on the y-axis.
2. Divide students into six groups. Have each group graph the trends across a different period on four separate transparency sheets. The teacher can graph period 1, as it contains only two elements. Overlay the transparencies to show trends within a period and to compare trends among the periods.

Name ______________________________ Date ____________

Activity 27: **Periodic Riddles**

Directions: Name each element based on the clues provided.

1. This element:
   (a) has only one energy level.
   (b) is part of an essential liquid.
   (c) is the lightest element.

   This element is ________________________.

2. This element:
   (a) has a smaller atomic radius than oxygen.
   (b) is one electron short of a full octet.
   (c) has the largest electronegativity.

   This element is ________________________.

3. This element:
   (a) has less ionization energy than sodium.
   (b) has a larger atomic radius than rubidium.
   (c) has a single letter symbol based on a Latin name.

   This element is ________________________.

4. This element:
   (a) has electrons in the *d* sublevel.
   (b) is the heaviest named element in its group.
   (c) is a shiny metal used for jewelry.

   This element is ________________________.

5. This element:
   (a) oxidizes faster than magnesium.
   (b) has a larger atomic radius than any of the actinides.
   (c) emits radiation.

   This element is ________________________.

(continued)

Name ______________________________ Date ____________

Activity 27: **Periodic Riddles** *(continued)*

6. This element:
   (a) is a nonmetal.
   (b) is the only gas in its family.
   (c) is used in dynamite and fertilizers.

This element is_______________________.

7. This element:
   (a) has an electronegativity greater than indium.
   (b) has an atomic radius larger than carbon.
   (c) has a first ionization energy greater than lead.
   (d) is used to make cans.

This element is _______________________.

8. This element:
   (a) has a first ionization energy greater than chromium.
   (b) has an atomic radius smaller than cobalt.
   (c) has an electronegativity greater than manganese.
   (d) is used to make steel and can be magnetized.

This element is _______________________.

9. This element:
   (a) is lighter than hafnium.
   (b) has a greater electronegativity than zirconium.
   (c) has a smaller atomic radius than scandium.
   (d) is used in aircraft manufacturing.

This element is _______________________.

10. This element:
    (a) has a larger atomic radius than sulfur.
    (b) has an electronegativity greater than tin.
    (c) has a first ionization energy greater than aluminum.
    (d) is a semimetal.
    (e) is used in computer chips.

This element is _______________________.

Activity 28a: **Periodic Table of Extraterrestrial Elements**

Objective: Students will receive information in stages about elements and use the new information to develop a periodic table of extraterrestrial elements. They will compare their periodic table with the periodic table of elements as modified by Moseley. Then, they will then use their periodic table to predict the characteristics of missing elements.

Materials: A copy of the three (physical, chemical, atomic mass) data tables on separate pieces of paper for each group. (Tables are given in activity 28b.)

Procedure:

1. Divide class into groups of 2-4 students.
2. Give each group a copy of the physical properties data table. Ask groups to develop a periodic table based upon the physical properties of the elements listed. They should write down an explanation for the order that they have assigned these elements.
3. As each group completes its periodic table and explanation, review the results to verify the group is proceeding in a logical manner. Once the first attempt is approved, give the group the chemical properties data table. Have the group modify its periodic table, if necessary. The groups will submit a second periodic table with justifications for changes or no changes.
4. Upon approval of the second submission and justification, give the group the final data table and questions. Using the data in this table, the group will adjust the order of the elements in its periodic table as necessary and justify any changes.
5. The final submission will consist of all three forms of the group's periodic table, all explanations and justifications, and the answers to the four questions.

Results: Students will have experienced the process used by Mendeleev and other scientists in developing the periodic table of elements. They will also be able to use their periodic table to predict missing elements and their properties.

Possible answers:

Phase A. Using only physical properties, students can divide the elements first by hardness, then by color, and finally by temperature.

Betelgeusean	Siriuson	Spican	Fomalhautian
Rigelian	Denebian	Vegan	Antarean
Procyon		Capellan	

(continued)

Phase B. With the addition of information on chemical properties, the groupings should change. Since Siriusian and Denebian have different chemical properties, they should be in different groups. Spican would also be in a separate group from Vegan and Capellan because Spican only reacts with oxygen. One example of a revised periodic table might look like:

Betelgeusean	Siriuson	Denebian	Vegan	Spican	Fomalhautian
Rigelian			Capellan		Antarean
Procyon					

Phase C. With the addition of relative atomic masses, the final version of their periodic table could be formatted so the elements increase in atomic mass from left to right and from top to bottom.

Fomalhautian	Spican				Betelgeusean
		Vegan	Siriuson		Rigelian
Antarean		Capellan		Denebian	Procyon

Phase D. QUESTIONS:

1. The two differences are: 1) the order is based on relative atomic masses (similar to Mendeleev's periodic table but different from Moseley's); and 2) the nonreactive elements are on the left, instead of the right side of the table.
2. The periodic table has three periods and six families.
3. There are at least fifteen extraterrestrial elements predicted by the number of the blank spaces in the periodic table. However, there may be more, if more periods are added.
4. The element in period 2, group 1, would be hard, pinkish-red, and nonreactive, with a melting point about 1525 K and a relative atomic mass of 4. The missing elements in group 2 would be hard, blue, have melting points greater than 1175 K, react only with oxygen, and have relative atomic masses of 5 (period 2) and 11 (period 3). The element in period 2, group 5 would be soft, gray, with a melting point less than 700 K, react with oxygen and acid, and have a relative atomic mass of 8. The element in period 3, group 4 would be soft, black, with a melting point greater than 1275 K, react with oxygen, acid, and water, and have a relative atomic mass of 13.

(continued)

Activity 28a: **Periodic Table of Extraterrestrial Elements** *(continued)*

Extensions:

1. Have students develop names and draw pictures of the missing elements.
2. Have students determine the phase of each element at room temperature.
3. Have students expand their periodic tables of extraterrestrial elements to include period 4.

Name ______________________________ Date ____________

Activity 28b: **Periodic Table of Extraterrestrial Elements**

Directions: Imagine you are a scientist aboard a spaceship. As you travel through the universe, you gather samples of different elements. You are to test these various samples and develop a periodic table with families and periods, similar to the periodic table chemists use on earth. Since there are several elements, you perform the testing in phases.

Phase A. The first data you collect are based on observations and tests for physical properties. The data are given in the following table. Devise a preliminary periodic table based on the data and justify your groupings.

PHYSICAL PROPERTIES

ELEMENT	COLOR	HARDNESS	MELTING POINT (K)
Antarean	Red	Hard	1575
Betelgeusean	Silvery, black	Soft	-25
Capellan	Turquoise	Hard	1275
Denebian	Gray	Soft	700
Fomalhautian	Pink	Hard	1475
Procyon	Silvery, black	Soft	175
Rigelian	Silvery, black	Soft	75
Siriuson	Black	Soft	575
Spican	Blue	Hard	1175
Vegan	Green	Soft	1275

(continued)

Name ______________________________ Date ____________

Activity 28b: **Periodic Table of Extraterrestrial Elements** *(continued)*

Phase B. In the second stage, you collect data on the chemical properties of the elements. Using the data listed below, modify your original periodic table and justify your new arrangement.

CHEMICAL PROPERTIES

ELEMENT	REACTS WITH WATER	REACTS WITH ACID	REACTS WITH OXYGEN	NO REACTION
Antarean				X
Betelgeusean	X		X	
Capellan		X	X	
Denebian		X	X	
Fomalhautian				X
Procyon	X		X	
Rigelian	X		X	
Siriuson	X	X	X	
Spican			X	
Vegan		X	X	

(continued)

Name ______________________________ Date ____________

Activity 28b: **Periodic Table of Extraterrestrial Elements** *(continued)*

Phase C. In the third stage, you determine the relative atomic masses for these extraterrestrial elements. Using the data, revise your periodic table and justify your new arrangement.

RELATIVE ATOMIC MASSES

ELEMENT	RELATIVE ATOMIC MASS	ELEMENT	RELATIVE ATOMIC MASS
Antarean	10	Procyon	15
Betelgeusean	3	Rigelian	9
Capellan	12	Siriuson	7
Denebian	14	Spican	2
Fomalhautian	1	Vegan	6

Phase D. Answer the following questions:

1. Explain two differences between the periodic table you have designed and the periodic table as modified by Moseley.

____

____

____

____

2. How many periods and how many families does your table have?

____

____

3. How many total extraterrestrial elements do you think exist?

____

____

4. Predict the properties of the missing elements.

____

____

____

____

Electron Configuration

Preliminary Handout:

- Electron Configuration (reproducible reading)

Activities:

29. Electron Configuration—Element to Electron Configuration (reproducible)
30. Electron Configuration—Electron Configuration to Element Name (reproducible)

31a. Element Yearbook (teacher guide page)

31b. Element Yearbook: Biographical Worksheet (reproducible)

Electron Configuration

The periodic table can be used to determine the electron configuration of an element. This is because each of its periods represents a principal electron energy level. The placement of the elements (except helium) reflects the energy sublevels in each electron energy level.

Groups 1 and 2 represent the *s* or **sharp sublevel**. The *s* sublevel has only one **orbital** and can contain up to two electrons. An orbital is visualized as a ring around the nucleus. This ring represents the area where the electron can, most probably, be found. Each orbital can contain zero, one, or two electrons. All group 1 elements have one electron in the *s* sublevel of the outermost energy level or **valence shell**. All group 2 elements have two electrons in the *s* sublevel of the valence shell. Hydrogen, which has only one electron, is placed at the top of group 1. Hydrogen is located on the periodic table in period 1, group 1. Therefore, the electron configuration for the hydrogen is $1s^1$. The first 1 represents the first energy level. The *s* represents the *s* sublevel. The superscript 1 represents hydrogen's one electron.

The first energy level has only one sublevel, the *s* sublevel. Therefore, it can contain only two electrons. Helium has only two electrons in the *s* sublevel. It should be placed at the top of group 2. However, helium is placed at the top of group 18 because its physical and chemical properties are similar to the rest of the elements in that group. Helium's electro configuration is $1s^2$. The first 1 represents the first energy level. The *s* represents the *s* sublevel. The superscript 2 represents helium's two electrons.

The first energy level is completely full with two electrons. This means helium has a stable configuration and is the first **noble gas** in Group 18. Group 18 elements are known as the noble gases because they are chemically inactive. Under normal conditions the noble gases do not react with any other element. Krypton, xenon, and radon have been forced to form compounds with fluorine under extreme conditions in a laboratory.

The second energy level, period 2, has two sublevels. These are the *s* and the *p* or **principal sublevel**. The *p* sublevel contains three orbitals. Since each orbital can contain two electrons, the *p* sublevel can contain up to six electrons. Groups 1 and 2 represent the *s* sublevel. Groups 13 through 18 represent the *p* sublevel. Lithium is in period 2, group 1 of the periodic table. Lithium's electron configuration is $1s^2s^1$. Lithium has three electrons. (The sum of the superscripts in the electron configuration equals the total number of electrons in the element.) The first two electrons fill the first energy level. The third electron enters the *s* sublevel of the second energy level. Electrons always fill the lowest energy levels first.

Beryllium is in period 2, group 2 of the periodic table. Beryllium's electron configuration is $1s^22s^2$. Beryllium has four electrons. The first two electrons fill the first energy level. The next two electrons enter the *s* sublevel of the second energy level. The *s* sublevel of the second energy level is now full, but there are six "spaces" in the *p* sublevel still open. The second energy level is not completely full until all eight "spaces" are filled with electrons. Therefore, beryllium is the first element in group 2 and not a noble gas.

(continued)

Boron is in period 2, group 13 of the period table. Boron's electron configuration is $1s^22s^22p^1$. Boron has five electrons. The first two electrons fill the first energy level. The next two electrons enter the *s* sublevel of the second energy level. The fifth electron enters the *p* sublevel of the second energy level. The rest of the elements of period 2 (carbon, nitrogen, oxygen, fluorine, and neon) continue to "place" electrons in the *p* sublevel of the second energy level. The superscript number following 2*p* increases with each additional electron. With neon, the second energy level is completely full. Neon's electron configuration is $1s^22s^22p^6$. Neon has ten electrons. (Neon's atomic number is 10, therefore neon has ten protons and ten electrons.) Neon is the second noble gas.

The third energy level has three sublevels: the *s*, the *p*, and the *d*, or **diffuse sublevel**. The *d* sublevel has five orbitals and can contain up to ten electrons. The *s* and *p* sublevels appear in period 3 of the periodic table of elements. The *d* sublevel appears in period 4. This is because the electrons in the *d* sublevel of the third energy level have higher energy than the electrons in the *s* sublevel of the fourth energy level, but lower energy than the electrons in the *p* sublevel of the fourth energy level. Period 4, therefore, contains elements having highest energy electrons in the 4*s*, 3*d*, and 4*p* sublevels. Period 5 contains elements having highest energy electrons: the 5*s*, 4*d*, and 5*p* sublevels.

The fourth energy level has four sublevels: the *s*, the *p*, the *d*, and the *f* or **fundamental sublevel**. The *f* sublevel has seven orbitals. It can contain up to fourteen electrons. The *f* sublevels for the fourth and fifth energy levels are located at the bottom of the periodic table. The elements in the **lanthanide series** have highest energy electrons in the 4*f* sublevel. The elements in the **actinide series** have highest energy electrons in the 5*f* sublevel. The lanthanide series is actually part of period 6. The actinide series is actually part of period 7. Electrons in the *f* sublevel of the fourth energy level have higher energy than electrons in the *s* sublevel of the sixth energy level. But they have lower energy than electrons in the *d* sublevel of the fifth energy level. Period 6, therefore, contains elements having highest energy electrons in the 6*s*, 4*f*, 5*d*, and 6*p* sublevels. Period 7 contains elements having highest energy electrons in the 7*s*, 5*f*, 6*d*, and 7*p* sublevels.

Henry Moseley placed the elements in the period table by atomic number. This arrangement also orders the elements by their number of electrons and electron energy. To determine the energy level for the *s* and *p* sublevels, use the periodic number on the left side of the periodic table. There is a tool to make it easier to determine the energy level for the *d* sublevel. Place a "3" between calcium and scandium, a "4" between strontium and yttrium, a "5" between barium and lanthanum, and a "6" between radium and actinium. You can also place a "4" between zinc and gallium, a "5" between cadmium and indium, a "6" between mercury and thallium, and a "7" between element 112 and the block for element 113. This will help you remember the energy level for the *p* sublevels. A "4*f*" can be written in front of the lanthanide series, and a "5*f*" can be written in front of the actinide series to remember the energy levels and *f* sublevel for these elements.

(continued)

Electron Configuration *(continued)*

To remember the sublevels, you can place an *s* above groups 1 and 2, a *d* above groups 3 through 12, and a *p* above groups 13–18. Students can also place a number with the letters to remember the superscript. For example, placing *d*1 over group 3 indicates that group 3 elements have one electron in the *d* sublevel.

Using the information offered in the periodic table, we can determine the electron configuration for every element. We can also determine the number of **valence electrons**. Valence electrons are the electrons found in the *s* and *p* sublevels of the highest principal energy level. For an example, the electron configuration of bromine is $1s^2 2s^2 s2p^6 3s^2 3p^6 4s^2 3d^{10} 4p^5$. Bromine has seven valence electrons. This is because its highest principal energy level (level 4) has two electrons in the *s* sublevel and five electrons in the *p* sublevel. The valence electrons determine the chemical properties of the element. Since all elements in a group have the same number of valence electrons, those elements also have similar properties. The only exceptions are hydrogen and helium. Hydrogen has one electron and, although located in group 1, does not react in the same manner as the rest of the group 1 elements. Helium has two electrons but is placed in group 18 because its physical and chemical properties are similar to the rest of the noble gases.

Name ______________________ Date __________

Activity 29: **Electron Configuration— Element to Electron Configuration**

Directions: Write the electron configurations for the following elements.

1. Li ______________________
2. Mg ______________________
3. O ______________________
4. Ar ______________________
5. Fe ______________________
6. Br ______________________
7. Hg ______________________
8. Si ______________________
9. U ______________________
10. K ______________________

Name ______________________ Date __________

Activity 30: **Electron Configuration— Electron Configuration to Element Name**

Directions: Write the names of the element with the following electron configurations.

1. $1s^1$ ____________________
2. $1s^2 2s^2 2p^1$ ____________________
3. $1s^2 2s^2 2p^5$ ____________________
4. $1s^2 2s^2 2p^6 3s^1$ ____________________
5. $1s^2 2s^2 2p^6 3s^2 3p^3$ ____________________
6. $1s^2 2s^2 2p^6 3s^2 3p^6 4s^2 3d^7$ ____________________
7. $1s^2 2s^2 2p^6 3s^2 3p^6 4s^2 3d^{10} 4p^6$ ____________________
8. $1s^2 2s^2 2p^6 3s^2 3p^6 4s^2 3d^{10} 4p^6 5s^2 4d^{10}$ ____________________
9. $1s^2 2s^2 2p^6 3s^2 3p^6 4s^2 3d^{10} 4p^6 5s^2 4d^{10} 5p^6 6s^2 4f^{14} 5d^{10} 6p^2$ ____________________
10. $1s^2 2s^2 2p^6 3s^2 3p^6 4s^2 3d^{10} 4p^6 5s^2 4d^{10} 5p^6 6s^2 4f^{14} 5d^{10} 6p^6 7s^2 5f^{11}$ ____________________

Activity 31a: **Element Yearbook**

Objective: Students will prepare an element "biography" and picture. The picture will resemble a yearbook picture, depicting characteristics of each element in human form.

Materials:
- Copy of worksheets for each student
- Resource materials to research elements

Procedure:
1. Have each student randomly select a different element.
2. Each student should complete a biography worksheet on the selected element, following the format outlined on the worksheet.
3. Using physical and chemical properties of the selected element, each student should draw a picture of the element depicting its special characteristics. The picture should use these properties as characteristics for a human representation. Each picture should be no larger than 17 cm × 20 cm. If it shows just the head and shoulders, it must be no smaller than 10 cm × 15 cm.
4. Compile all of the students' biography worksheets and pictures to form an element yearbook.

Results: Students should use the available resources to complete the biography worksheet. Utilizing their creative abilities, they should draw a picture of the element that shows the various chemical and physical properties of the element. The completed element yearbook can then be used as a source of information for the class.

Name ______________________________ Date ____________

Activity 31b: **Element Yearbook: Biographical Worksheet**

Element name: ______________________ Element symbol: ______________

Atomic number: ______________ Group name or number: ______________

Atomic mass (to nearest thousandth): ______________________________

Number of protons: ______________ Number of neutrons: ______________

Number of electrons: ____________ Number of valence electrons: ____________

Electron configuration:

____

____

HISTORY

Discovered by: ____

Year discovered: ________ Where discovered: ______________________

Derivation of name/symbol: ______________________________________

PHYSICAL AND CHEMICAL CHARACTERISTICS

Phase at room temperature: ______________________________________

Density at room temperature (g/cm^3): ____________________________

Melting point (K): ________ Boiling point (K) ______________________

Color: ______________________ Odor: ______________________

Oxidation states: ______________ Ionic or covalent bond: ______________

Electronegativity (Pauling): ______________ Electron affinity (kJ/mol): ________

Reactivity with oxygen, water, acids or bases:

____

____

(continued)

Name ______________________ Date ____________

Activity 31b: **Element Yearbook: Biographical Worksheet** *(continued)*

CURRENT INFORMATION

Where found (specific minerals or sources/specific countries):

____

____

____

____

Uses:

____

____

____

____

Toxicity/hazards:

____

____

____

____

Abundance:

____

____

____

____

Groups

Preliminary Handout:

- Groups (reproducible reading)

Activities:

32a. Group Codes (teacher guide page)

32b. Group Codes (reproducible)

33. Groups 4–11 Crossword (reproducible)

34. Groups 12–16 Word Game (reproducible)

Groups

The vertical columns of the periodic table are known as the groups or families. There are eighteen groups numbered from 1 to 18 starting on the left side of the periodic table. These elements are grouped together since they have similar chemical properties. The chemical properties are based on the electron configuration—specifically, the valence electrons (the electrons in the highest or outermost energy level).

The elements with one or two valence electrons are metals. They are unstable, and form positive ions (or cations). These elements will lose electrons in order to reach the stable configuration of the noble gases.

The elements with seven valence electrons, except astatine, are nonmetals. Also unstable, they form negative ions (or anions). They will gain electrons in order to reach the stable configuration of the noble gases. The noble gases have eight valence electrons (except helium) and tend to be nonreactive. This nonreactivity results from their electron configuration: eight valence electrons completely fill the *s* and *p* orbitals of the valence sublevels. Helium is an exception, as it only requires two valence electrons to fill the 1*s* sublevel. At one time, chemists called these gases "inert"—that is, nonreactive—because they believed these gases did not react with other elements. However, the name has been changed, since chemists have now been able to react xenon, krypton, and radon with fluorine, the most active nonmetal.

The major groups of the periodic table of elements are the alkali metals, the alkaline earth metals, halogens, and noble gases (groups 1, 2, 17 and 18, respectively). These are described in more detail in later pages.

Groups 3 through 11 are known as the transition elements. These elements have several characteristics in common. They are all metals. Most have more than one oxidation state and form colored ions. However, the elements in each group have their own distinctive characteristics. For example, Group 11 consists of copper, silver, gold and unununium. The first three elements are soft, shiny metals that are malleable and very ductile. They can be pounded into thin sheets and drawn into very fine wires. Group 3 includes scandium, yttrium, lanthanum and actinium. The lanthanide and actinide series appear to be part of group 3 although they are placed in two rows outside the main body of the periodic table. Based on Glenn Seaborg's actinide concept, the lanthanide and actinide series were placed outside the main body of the periodic table. This placement helped to show the similarities in the lanthanide and actinide series and their relationship to the other elements. Since they have characteristics similar to lanthanum and actinium respectively, they are described in more detail later.

The other transition groups are listed below:

- Group 4—titanium, zirconium, hafnium, and rutherfordium
- Group 5—vanadium, niobium, tantalum, and dubnium
- Group 6—chromium, molybdenum, tungsten, and seaborgium
- Group 7—manganese, technetium, rhenium, and bohrium
- Group 8—iron, ruthenium, osmium, and hassium
- Group 9—cobalt, rhodium, iridium, and mendelevium
- Group 10—nickel, palladium, platinum, and ununnilium

The heaviest member of each group is radioactive and has a very short half-life.

(continued)

Group 12 does not fit the International Union of Pure and Applied Chemistry (IUPAC) definition of transition elements. It is, however, a short group and is usually discussed with the transition elements. Group 12 consists of zinc, cadmium, mercury, and ununbium. Some interesting characteristics of this group include:

- Mercury is a liquid; the other members of the group are solids.
- Ununbium is radioactive.
- Both mercury and cadmium are toxic. However, mercury is still used in barometers and some thermometers. Cadmium is still found in some rechargeable batteries.
- Zinc is one of several essential elements for human health. It is also used with copper to form the alloy brass.

Group 13 currently has five members. They are boron, aluminum, gallium, indium, and thallium. Characteristics of this group include:

- Boron is a semimetal, while the other members are metals.
- All these elements have a 3+ charge. Therefore, they have three valence electrons that they tend to lose when forming cations.
- Boron is extracted from the ore borax, and is used in detergents and fire retardants.
- Aluminum is a lightweight metal that has many uses. One of the most common uses is in aluminum soda cans.
- Gallium and indium are also frequently used, although they are not as easily recognizable as aluminum. Gallium is used in microwave equipment. Indium is used in safety devices such as fire-door links, fusible plugs, and sprinkler heads.
- Thallium is the only toxic member of the family, and has very limited use.

Group 14 consists of carbon, silicon, germanium, tin, lead, and ununquadium. Characteristics of this group include:

- Carbon is a nonmetal; silicon and germanium are semimetals; and the other members of the family are metals.
- Carbon is probably the most important member of this group. This is because of its ability to form multiple covalent bonds with itself and other elements. Elemental carbon is found as graphite, coal, or diamonds. The study of carbon and its compounds forms a separate branch of chemistry known as organic chemistry.
- Silicon is used as a semiconductor in computer chips. It has given its name to an area in California known for its computer and electronics industry: Silicon Valley. Germanium is also used in semiconductors and for infrared devices.

(continued)

- Tin is a stronger element than aluminum, and is used to make tin cans. It also was used in the Tin Man from *The Wizard of Oz.*
- Lead is a heavy metal. It is highly toxic, and so we no longer add lead to paint and gasoline.
- Ununquadium is the newest member of the group. It has a very short half-life.

Group 15 is sometimes called the nitrogen family. It currently has five members: nitrogen, phosphorus, arsenic, antimony, and bismuth. Characteristics of this group include:

- Nitrogen and phosphorus are nonmetals; arsenic and antimony are semimetals; and bismuth is a radioactive metal. These elements all have five electrons in the valence sublevel and form ions with a 3- charge.
- Nitrogen and phosphorus are both essential for life. They are both found in fertilizers.
- Nitrogen is a component of explosives. Phosphorus has been used in detergents.
- One form of phosphorus, white phosphorus, is very flammable.
- Although both arsenic and antimony are toxic, only arsenic is used as a pesticide.
- The heaviest member of this group, bismuth, is used in cosmetics and pharmaceuticals.

Group 16 is sometimes called the oxygen family. It consists of oxygen, sulfur, selenium, tellurium, polonium, and ununhexium. Characteristics of this group include:

- Oxygen, sulfur, and selenium are nonmetals; tellurium is a semimetal; and polonium and ununhexium are radioactive metals. These elements have 6 valence electrons and have an apparent charge of 2-.
- Oxygen and sulfur are essential to life.
- Sulfur is distinctive because in its elemental form it is yellow. When burnt, it smells like rotten eggs.
- Selenium and tellurium are slightly toxic. However, selenium is used in photoelectric and solar cells. Tellurium is used in electronics and catalysts.
- Polonium and ununhexium are the radioactive members of the family. Polonium was discovered in 1898 by Marie Curie. Ununhexium was synthesized in 1999.

Activity 32a: **Group Codes**

Objective: Students will be able to place elements into the major groups of the periodic table.

Materials:

- Copy of the periodic table for each student (A key should be added to the periodic table in order to identify the families.)
- Koosh with list of element names or beach ball with element symbols on it (If using a beach ball, write the element symbols for elements in groups 1, 2, 17, and 18 randomly on the beach ball. If using a koosh, compile a random listing of the element names for the elements in groups 1, 2, 17, and 18.)
- Copy of the worksheet for each student

Procedure:

1. Have students draw symbols or pictures to represent different groups (families) in each element block for groups 1, 2, 17, and 18. If the symbols or pictures used represent a specific property, they can also help students remember at least one distinguishing characteristic for the family. Examples of pictures that can be used are:
   a. Alkali metals can be represented by flames to indicate they react explosively with water.
   b. Alkaline earth metals can be represented by a circle to represent the earth.
   c. Halogens can be represented by a hand (since they "grab" electrons) or salt shakers (since halogen is from the Greek words meaning "salt former").
   d. Noble gases can be represented by crowns, since they do not react or "associate" with other elements.
2. Toss the koosh or beach ball to a student. If using a koosh, the student catching the koosh must state the family or group name for the first element on the previously compiled list. If using the beach ball, the student must state the family or group name for the element symbol under his/her left thumb. Continue until all students have identified at least one family and all elements in the four groups have been covered.
3. Have students complete worksheet.

Results: Students should be able to place the elements in the correct family or group. If the symbols or pictures are representative, they should also be able to name at least one distinguishing characteristic of the family.

Name ______________________________ Date ____________

Activity 32b: **Group Codes**

Directions: Complete the chart below by indicating the group number for the following elements.

ELEMENT	GROUP NUMBER
Potassium	
Fluorine	
Silicon	
Radon	
Gold	
Beryllium	
Oxygen	
Neon	
Rubidium	
Tungsten	
Arsenic	
Calcium	
Sodium	
Aluminum	
Chromium	
Iodine	
Helium	
Nitrogen	
Iron	
Zinc	

Name ______________________ Date __________

Activity 33: **Groups 4–11 Crossword**

Directions: Write the element name and group number next to its symbol. Then write the element name in its appropriate place in the crossword puzzle.

ACROSS

3. ______________(V)
5. ______________(Bh)
7. ______________(Os)
9. ______________(Ta)
11. ______________(Mt)
13. ______________(Re)
15. ______________(Mn)
16. ______________(Co)
17. ______________(Ti)
19. ______________(Ni)

DOWN

1. ______________(Pd)
2. ______________(Mo)
4. ______________(Pt)
6. ______________(Hf)
8. ______________(Ag)
10. ______________(Tc)
12. ______________(W)
14. ______________(Nb)
16. ______________(Cu)
18. ______________(Fe)

Name ____________________ Date ________

Activity 34: **Groups 12–16 Word Game**

Directions: Identify the element and the group to which it belongs from the statements and the letter clues.

1. ___ ___ L ___ ___ ___ ___ is used to make chips for computers. Group _____
2. ___ ___ T ___ ___ ___ ___ ___ is an important ingredient used in plant fertilizers. Group _____
3. ___ X ___ ___ ___ ___ is a component of air that is necessary for life. Group _____
4. B ___ ___ ___ ___ is extracted from borax and is used in fire retardants and detergents. Group _____
5. ___ ___ ___ ___ ___ ___ Y is a liquid metal used in barometers. Group _____
6. ___ ___ L L ___ ___ ___ is used in microwave equipment. Group _____
7. When an automobile is not properly in tune, it will not burn the fuel properly, causing ___ ___ R ___ ___ ___ deposits to build up on the engine. Group _____
8. ___ ___ ___ F ___ ___ is a yellow nonmetal that smells of rotten eggs when burnt. Group _____
9. ___ ___ ___ M ___ ___ ___ ___ soda cans are easily crushed. Group _____
10. When combined with copper, ___ ___ ___ C forms the alloy brass. Group _____
11. This element is abundant in the soil, and is an effective poison. A famous play was titled ___ R S ___ ___ ___ ___ *and Old Lace*. Group _____
12. ___ E ___ ___ N ___ ___ ___ is used in photoelectric and solar cells. Group _____
13. ___ ___ D M ___ ___ ___ is used in rechargeable batteries, but is toxic. Group _____
14. Some foods are canned commercially in ___ I ___ cans. The inside of these cans is coated with a protective finish to prevent corrosion. Group _____
15. P ___ ___ ___ N ___ ___ ___ is a radioactive element named after Poland. Group _____
16. ___ ___ ___ M ___ ___ H is found in pharmaceuticals and cosmetics. Group _____
17. The white form of ___ ___ ___ ___ P ___ ___ ___ ___ S is highly flammable. Group _____
18. Diamonds are a valuable form of ___ ___ ___ B ___ ___. Group _____
19. The ___ ___ N Man in *The Wizard of Oz* wanted a heart. Group _____
20. ___ N ___ ___ ___ ___ is used in safety devices. Group _____

Alkali Metals

Preliminary Handout:

- Alkali Metals (reproducible reading)

Activities:

35. Alkali Metal Clues (reproducible)
36. Alkali Metals Quiz (reproducible)

Alkali Metals

The alkali metals are lithium, sodium, potassium, rubidium, cesium, and francium. Characteristics of the alkali metals include:

- They have one electron in the outermost energy level. These are very active metals and easily lose their single valence electron.
- The alkali metals are very soft and are easily cut with a knife.
- They react readily with oxygen, and when exposed to air quickly lose their luster.
- They react with water to generate hydrogen gas, and therefore form very strong bases.
- The reactivity of these elements increases as their atomic number increases, with lithium being the least reactive and francium being the most reactive. This increase in reactivity can be seen in the elapsed time it takes for the element to lose its luster or in the strength of its reaction with water. If pieces of each element are sliced to expose a lustrous surface, rubidium will tarnish as soon as it is sliced, whereas lithium will remain lustrous for a few minutes.
- Each alkali metal reacts differently if very small pieces of each element are placed in a container of water. Lithium will skim across the surface, releasing hydrogen gas. Sodium will skim across the surface, and the escaping hydrogen gas may burst into flames. Potassium releases hydrogen gas so rapidly that the piece of potassium appears to burst into flames immediately. Rubidium causes an explosive reaction.

Name ______________________________ Date ____________

Activity 35: **Alkali Metal Clues**

Directions: List the six alkali metals, their symbols and atomic numbers. Use the spaces and letter clues to answer the riddles.

ELEMENT	SYMBOL	ATOMIC NUMBER

1. This element is used in antidepressant medicines.

   __ __ __ H __ __ __

2. This element is the only radioactive element in this group.

   __ __ __ __ C __ __ __

3. This element is used both as a replacement in low-sodium diets and to make stronger glass.

   __ __ T __ __ __ __ __ __

4. This element is a liquid at 30°C.

   __ E __ __ __ __

5. This element is a component of common table salt.

   __ __ D __ __ __

6. This solid element reacts so violently with water that it causes an explosion.

   __ __ B __ __ __ __ __

Name ________________________________ Date ____________

Activity 36: **Alkali Metals Quiz**

Directions: Fill in the blanks with the correct answers.

1. The elements in group ______________ are known as the alkali metals.
2. They all have ______________ valence electron(s) and form ions with a ______________ charge.
3. These elements form a positive ion known as a ______________.
4. The element with the lowest atomic mass in this group is ______________ and its symbol is ______________.
5. The element with the highest atomic mass in this group is ______________ and its symbol is ______________.
6. ______________ is a liquid at 30°C. ______________ is radioactive.
7. The first four elements of the group (______________, ______________, ______________, ______________) are soft, solid metals that can be cut with a ______________.
8. When exposed to air, these elements tarnish, indicating a reaction with ______________.
9. These elements react with water to release ______________ gas.
10. These metals form strong ______________.

Alkaline Earth Metals

Preliminary Handouts:
- Alkaline Earth Metals (reproducible reading)

Activities:
37. Alkaline Earth Metals, Sources, and Uses (reproducible)
38. Alkaline Earth Metals Puzzle (reproducible)

Alkaline Earth Metals

The alkaline earth metals are beryllium, magnesium, calcium, strontium, barium, and radium. They have two electrons in the outermost electron energy level, are highly metallic, good conductors of electricity and form alkaline solutions with water. They are not found free in nature, but can be extracted from their combined states. All the alkaline earth metals except beryllium react with air and water to varying degrees. They were originally called "earths" because they were thought to be nonmetallic, insoluble in water, and unchanged by fire. It was later discovered that these "elements" were, in fact, the oxides of the alkaline earth metals and not the **free or uncombined elements**.

Characteristics of the alkaline earth metals include:

- Beryllium is the lightest of the alkaline earth metals. It is extracted from the minerals beryl and chrysoberyl. The mineral beryl is also a source of aquamarines and emeralds. Beryllium is a poison and may be carcinogenic. It is alloyed with copper to make spark-proof tools.
- Magnesium is extracted from seawater or the mineral dolomite. It burns with a very bright light when ignited in air. Magnesium is used as a sacrificial electrode to protect other metals. It is an essential element for humans, and is used in laxatives (milk of magnesia), Epsom salts, and dyes.
- Calcium is another element essential for humans. It is found in the minerals calcite (limestone, marble, and chalk), dolomite, and gypsum. Calcium is used in the manufacture of rare earth metals, zirconium, thorium, and uranium. It is also used in drying agents and cement.
- Strontium is found in the minerals celestite and strontianite. It produces red coloring in flares and fireworks. It is also used in the special glass needed for televisions.
- Barium is extracted from the mineral barite. It is used in drilling fluids, paint pigments and for body imaging. Patients drink barium sulfate so doctors can see their gastrointestinal tracts with the help of x-ray machines.
- Radium is the heaviest of the alkaline earth metals, and is the only radioactive member of the family. It is found in the uranium ore pitchblende. It was used as a luminous paint for watch dials and to treat cancer. Both of these uses have been discontinued due to the hazards of radiation exposure.

Name ______________________________ Date ____________

Activity 37: **Alkaline Earth Metals: Sources and Uses**

Directions: Complete the table by writing the correct element name next to its source and uses.

ELEMENT	SOURCE	USES
	Strontianite	Special glass for TVs
	Seawater	Sacrificial electrodes
	Beryl	Spark-proof tools
	Dolomite	Manufacture of rare earth metals
	Barite	Drilling fluids and body imaging
	Celestite	Red flares and fireworks
	Calcite	Drying agent and cement
	Dolomite	Laxatives, dyes and Epsom salts
	Pitchblende	Cancer treatment and watch dials

Name ______________________________ Date ______________

Activity 38: **Alkaline Earth Metals Puzzle**

Directions: In the chart below, write the names of all the alkaline earth metals, their symbols, and their atomic numbers. Then, in the element puzzle box, highlight the symbols and atomic numbers for all the alkaline earth metals. The highlighted shape will provide you with the answer to this incomplete sentence:

All alkaline earth metals form ions with a ______________ charge.

ELEMENT	SYMBOL	ATOMIC NUMBER

ELEMENT PUZZLE BOX

Bi	Co	18	Sc	Ni	43	82	Rb	Cs	87
Md	24	Be	Mg	Mo	42	41	56	Ge	79
Cu	88	Mn	29	4	Na	6	20	15	84
36	Mo	Fe	Cu	Ca	Ru	Ca	Ra	Ba	Xe
85	Sc	17	As	Ba	77	Tl	88	57	55
52	34	Au	Mg	37	76	Hg	12	10	Ta
Fe	He	Sr	Ag	51	82	19	13	Pb	Re
Hf	4	38	20	Sr	Sb	Co	Ga	Li	Pt
Rn	21	3	Ti	Sn	Pb	Eu	16	34	Pd
22	Ti	5	Zr	Ti	Ta	Sm	40	54	Kr

Halogens

Preliminary Handout:
- Halogens (reproducible reading)

Activities:
39. Halogen Color Codes (reproducible)
40. Halogens Quiz (reproducible)

Halogens

The halogens are fluorine, chlorine, bromine, iodine, and astatine. The name halogen comes from the Greek words meaning "salt former." The name is aptly given because halogens readily form salts with alkali metals. The halogens have seven valence electrons and tend to take electrons. Thus, they form negative ions or anions with a 1- charge.

Characteristics of the halogens include:

- Fluorine, chlorine, bromine, and iodine are the most reactive nonmetals, and are not found free in nature. In fact, in their elemental form, when not combined with other elements, they form diatomic molecules.
- The gaseous phase of the four nonmetallic halogens have distinct colors. Fluorine is a pale yellow gas, chlorine is a yellow green gas, bromine forms a red gas, and iodine sublimes into a purple gas. All the gases are toxic and corrosive.
- Astatine is one of the semimetals, and is radioactive. Astatine is rarely found in nature; it is usually made in research laboratories by combining bismuth and an alpha particle.

Name ______________________________ Date ____________

Activity 39: **Halogen Color Codes**

Directions: The nonmetallic halogens form distinctly colored gases. Write the names of these elements, their symbols, and their atomic numbers in the color of their gases in the table below. For the semimetal member, use black. Then, using the color clues in parentheses, answer the questions.

ELEMENT	SYMBOL	ATOMIC NUMBER

1. This element is the liquid member of the family. (red) ____________________
2. This element is used in photography and as an antiseptic. (violet)

   ____
3. All of the elements are nonmetals except this member. (black)

   ____
4. This element is the most reactive nonmetal. (yellow) ____________________
5. This element is extracted from sea water. (red) ____________________
6. This element forms a poisonous gas and is used in pools. (green)

   ____
7. This element is the radioactive member of the family. (black)

   ____
8. This element combines with sodium to form common table salt. (green)

   ____
9. This element is a solid nonmetal. (violet) ____________________
10. This is the lightest member of the group. (yellow) ____________________

Name ______________________________ Date ____________

Activity 40: **Halogens Quiz**

Part A. Directions: Circle the correct response from the choices given in parentheses.

1. The halogens are found in Group (**17** or **18**).
2. The halogens form (**anions** or **cations**) with a (**1+** or **1-**) charge.
3. The first four halogens are diatomic (**metals**, **semimetals**, or **nonmetals**).
4. The element with the highest electronegativity is (**fluorine**, **chlorine**, **bromine**, **iodine**, or **astatine**).
5. These elements form strong (**acids** or **bases**) when mixed with water.

Part B. Directions: Indicate if the statement is true or false. If false, change the underlined word(s) to make the statement true.

6. <u>Astatine</u> is a radioactive product of uranium decay. ____________________
7. All of the nonmetallic halogens form <u>non-corrosive</u> gases. ____________________
8. Although the elemental forms are toxic, the ions of <u>fluorine</u>, <u>chlorine</u>, and <u>iodine</u> are essential for humans. ____________________

Noble Gases

Preliminary Handout:

- Noble Gases (reproducible reading)

Activities:

41. Noble Gases Puzzle (reproducible)
42. Noble Gases Pictionary® (reproducible)

Noble Gases

The noble gases are helium, neon, argon, krypton, xenon, and radon. They are the least reactive of the elements because of their electron configuration. All of these elements, except helium, have eight electrons in their outermost energy level and are, therefore, complete. Although helium only has two electrons in its outermost energy level, it is complete since the 1*s* sublevel can only hold two electrons. Since the valence shells of the noble gases are complete, they do not react readily with any other elements. Only the heavier noble gases (krypton, xenon, and radon) have reacted with fluorine under controlled laboratory conditions.

Characteristics of the noble gases include:

- All of the noble gases are colorless, odorless gases.
- All except radon are nontoxic.
- Radon is radioactive and emits alpha particles. It is formed when uranium and thorium disintegrate, and also collects over radium samples. This noble gas phosphoresces when cooled below its freezing point.
- Xenon is the proverbial "lead balloon." When a balloon is filled with xenon, it falls with a thud.
- Krypton also is heavier than air and a balloon filled with krypton will fall toward the ground. The red-orange line in krypton's spectrum is used as the fundamental standard for the meter. Krypton is also used in light bulbs and welding.
- Argon is probably best known as a welding gas. It is also used in light bulbs, including mercury vapor lamps for street lights.
- Neon is probably best known as the gas in neon signs. It is also used in welding and is mixed with oxygen for deep sea divers.
- Helium is the lightest of the noble gases and is best known as the gas used for party balloons or dirigibles. It is also used in mercury vapor lamps and mixed with oxygen for deep sea divers.

All of the noble gases except radon have names derived from Greek.

- Helium comes from *helos* meaning "sun" since helium was first observed in the sun's spectrum.
- *Neon* is the Greek word meaning "new."
- Argon is derived from *argos* meaning "inactive" since it does not react with any other element.
- Krypton comes from *kryptos* meaning hidden.
- The name xenon is derived from the Greek word *xenos* meaning "stranger."

Name ______________________________ Date ____________

Activity 41: **Noble Gases Puzzle**

Directions: Two words once used by scientists to describe the noble gases no longer apply. What are they? Use the names of the six noble gases to find those words.

First, use the letter clues in each horizontal row to complete the name of the correct noble gas. You will use most names more than once—in fact, you may use one noble gas name three times! When you have completed the puzzle, read down vertically starting at the black arrow. Now you know the answer to the question asked above!

Name ______________________________ Date ____________

Activity 42: **Noble Gases Pictionary®**

Directions: List the noble gases, their symbols, and their atomic numbers in the chart below. Then use the picture clues to identify each noble gas.

ELEMENT	SYMBOL	ATOMIC NUMBER

1

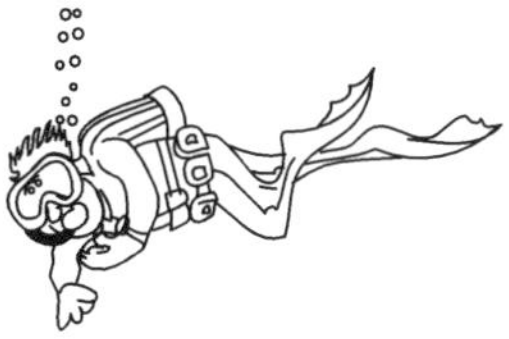

2

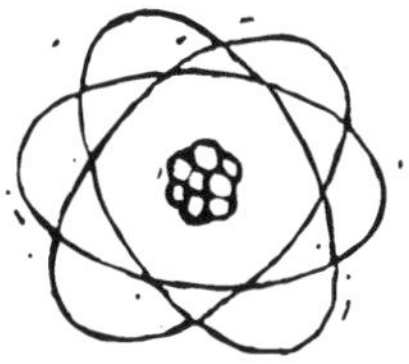

3

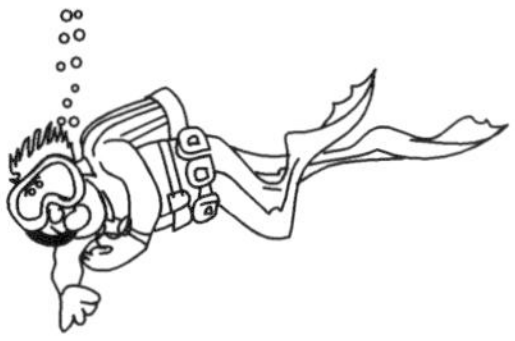

(continued)

Name ______________________________ Date ____________

Activity 42: **Noble Gases Pictionary®** *(continued)*

4

5

6

Transition Elements

Preliminary Handout:
- Transition Elements (reproducible reading)

Activities:
43. Transition Elements Crossword (reproducible)
44. Transition Element Group Numbers (reproducible)
45. Find the Series (reproducible)
46. Actinide Series Word Find (reproducible)

Transition Elements

The transition elements are all metals and usually exhibit two or more oxidation states. Transition elements, as defined by the International Union of Pure and Applied Chemistry (IUPAC), are elements with atoms having an incomplete *d* sublevel or atoms that form cations having an incomplete *d* sublevel.

According to this definition, the transition elements are:

- Period 4—scandium, titanium, vanadium, chromium, manganese, iron, cobalt, nickel, and copper
- Period 5—yttrium, zirconium, niobium, molybdenum, technetium, ruthenium, rhodium, palladium, and silver
- Period 6—lanthanum, hafnium, tantalum, tungsten, rhenium, osmium, iridium, platinum, and gold
- Period 7 (the new synthetic elements)—actinium, rutherfordium, dubnium, seaborgium, bohrium, hassium, meitnerium, 110, and 111.

The elements in Group 12 (zinc, cadmium, mercury, and 112) have complete *d* sublevels and, therefore, by definition, are not transition elements.

Outside the main body of the periodic table are two rows of elements representing the filling of the *f* sublevel. Within Period 6 is a series of elements known as the lanthanides. These transition elements are characterized by the successive filling of the 4*f* sublevel. These elements are cerium, praseodymium, neodymium, promethium, samarium, europium, gadolinium, terbium, dysprosium, holmium, erbium, thulium, ytterbium, and lutetium. In some cases lanthanum is included in the series and lutetium becomes the first element in the Period 6 transition elements belonging to Group 3.

The lanthanides have been called rare earth elements. However, this grouping includes scandium, yttrium, and lanthanum, which are chemically similar to the other members of this family. Although these elements are called rare earth elements, they are not, in fact, rare; they are found in small concentrations in many places around the earth. They are generally mixed with other rare earth elements. When separated and purified, the pure metal is bright and silvery.

Within Period 7 is a series of elements known as the actinides. These transition elements are characterized by the successive filling of the 5*f* sublevel. These elements are thorium, protactinium, uranium, neptunium, plutonium, americium, curium, berkelium, californium, einsteinium, fermium, mendelevium, nobelium, and lawrencium. In some cases, actinium is included in the series. Lawrencium then becomes the first of the period 7 transition elements belonging to group 3. The actinides are all radioactive, most of which are synthetic.

Name ______________________________ Date ______________

Activity 43: **Transition Elements Crossword**

Directions: Use the chemical symbols of the transition elements to determine each element name. Write the element name on the line by its chemical symbol. Then write the name in the puzzle.

ACROSS

1. ____________________(Co)
6. ____________________(W)
7. ____________________(Rh)
8. ____________________(Ag)
11. ____________________(Cr)
14. ____________________(Bh)
16. ____________________(Er)
18. ____________________(Fm)
19. ____________________(Y)
20. ____________________(Dy)
21. ____________________(Ni)

DOWN

1. ____________________(Cu)
2. ____________________(La)
3. ____________________(Os)
4. ____________________(Re)
5. ____________________(Au)
9. ____________________(Tc)
10. ____________________(Ho)
12. ____________________(Hf)
13. ____________________(Fe)
15. ____________________(Ir)
17. ____________________(Mo)

Name ______________________ Date __________

Activity 44: **Transition Element Group Numbers**

Directions: Identify the group to which each transition element belongs. Use numbers 3 through 12 to complete the table. Group 12, although not transition elements by the IUPAC definition, will be included. The lanthanide and actinide series will be excluded.

ELEMENT	GROUP	ELEMENT	GROUP	ELEMENT	GROUP
Bohrium		Manganese		Scandium	
Cadmium		Meitnerium		Seaborgium	
Cobalt		Mercury		Silver	
Copper		Molybdenum		Tantalum	
Chromium		Nickel		Technetium	
Dubnium		Niobium		Titanium	
Gold		Osmium		Tungsten	
Hafnium		Palladium		Vanadium	
Hassium		Platinum		Yttrium	
Iridium		Rhenium		Zinc	
Iron		Rhodium		Zirconium	
Lawrencium		Ruthenium			

Name ______________________________ Date ______________

Activity 45: **Find the Series**

Directions: This series of elements is between the atomic numbers of ____ and ____. It is called the ____________ series. Cross out the symbols of the elements that do not belong to this series for a letter clue. Then list the names of the elements and their symbols in the chart below. Hint: There are only two series, the lanthanide series and the actinide series.

Cf	Ge	Ni	As	Sb	Sn	Sg
At	Er	Xe	Au	Hf	Re	Ac
Te	Yb	Ag	Si	Br	Ar	No
Cl	Tm	Mg	Nb	Mo	Os	Es
Br	La	Ta	He	Rh	Ti	Am
Kr	Ce	Cs	Ne	Ta	Ir	Zr
Rn	Pr	Pb	Bi	Po	Br	Ga
Xe	Nd	Li	Na	Mn	Ca	Be
Pt	Pm	Ra	Lr	Lu	Tc	Th
Rb	Sm	Cu	Sc	Cd	Hg	Pu
Sr	Eu	Co	Ba	Zn	Ru	Np
Pa	Gd	Tb	Dy	Ho	Lu	Si
Cr	Tl	Nb	Pd	Bk	Md	Al

ELEMENT	SYMBOL

Name ______________________________ Date ____________

Activity 46: **Actinide Series Word Find**

Directions: List the names of the elements in the actinide series, including actinium, in the following chart. Then find and circle each element on the word find puzzle that follows.

P	O	N	E	P	T	U	N	I	U	M	P	P	L	K	J	H	G	F
M	N	B	V	C	X	Z	A	S	U	Y	T	R	Q	W	E	R	T	Y
Q	W	E	F	E	R	M	J	I	U	M	P	O	J	H	G	H	D	S
A	Q	Z	B	E	F	C	N	A	R	O	H	T	H	O	R	I	U	M
P	C	A	L	I	F	O	R	N	I	U	M	A	A	A	C	V	B	N
Y	U	R	A	D	T	D	A	G	O	U	R	C	S	E	M	Y	H	N
X	R	E	M	U	I	U	N	E	I	O	G	T	S	S	U	I	M	M
N	I	W	L	M	E	N	D	V	A	N	X	I	L	U	I	A	U	L
O	U	P	B	E	R	K	E	L	I	U	M	N	O	B	C	I	I	O
B	M	V	R	A	G	L	A	W	R	C	Q	I	Q	Y	N	I	N	X
E	I	N	S	T	E	I	N	I	U	M	R	U	C	I	E	M	A	Z
L	Z	X	Y	D	P	L	U	T	O	U	I	M	T	B	R	C	R	A
I	P	O	N	C	A	L	I	F	O	I	D	C	U	R	W	G	U	I
U	H	E	M	U	I	C	I	R	E	M	A	I	L	R	A	H	G	F
M	M	Q	U	E	E	T	Y	U	I	R	P	O	I	W	L	T	R	E
Z	X	C	V	B	N	M	L	K	J	E	Q	W	E	L	T	Y	U	I
T	Y	U	R	W	A	L	K	J	H	F	Z	X	C	V	B	N	M	L

Synthetic Elements

Preliminary Handout:

- Synthetic Elements (reproducible reading)

Activities:

47. Synthetic Elements Puzzle (reproducible)
48. Transuranium Elements Match (reproducible)

Synthetic Elements

Technetium and all of the transuranium elements (elements with atomic numbers greater than 92), except plutonium, are synthetic. The rest of the elements are considered natural elements since they are found in nature (though sometimes only in very minute quantities). Plutonium is an element that is found in nature in extremely small amounts. Since we use plutonium as a nuclear fuel, usable quantities of plutonium are all synthetic.

The transuranium elements are created by one of two processes. In the first method, the nucleus of an atom is bombarded with slow neutrons. The nucleus gains a neutron with the release of gamma radiation during the bombardment. Then, negative beta particle decay occurs spontaneously. Negative beta particle decay results when a neutral neutron is transformed into a positive proton; that transformation releases a negative beta particle. Since the beta particle has negligible mass, the mass number of the nucleus does not change. However, the result is an additional proton in the nucleus, which means that the atom has become a different element. This method is frequently used to produce neptunium from uranium. In the first step, uranium 238 gains a neutron to become uranium 239. In the beta particle decay, a neutron becomes a proton changing uranium into neptunium; uranium 239 becomes neptunium 239. Because of the negligible mass loss, the mass number of the new element remains 239.

Transuranium elements may also be manufactured by bombarding the nucleus of an atom with charged particles fired from an accelerator. The alpha particle, or helium nucleus, is a small ion; it can be easily accelerated and incorporated into a new element. However, to create the heavier and, it is hoped, more stable transuranium elements, larger ions must be used for bombardment. This requires the use of heavy-ion linear accelerators that consist of two accelerators. One accelerator removes many of the atom's electrons to produce more highly charged ions. Next, the second accelerator accelerates these ions to about one-seventh the speed of light. These ions can then be used to bombard the nuclei of other atoms. When the ions combine with other atoms, they form new, heavier elements.

Name ______________________________ Date __________

Activity 47: **Synthetic Elements Puzzle**

Directions: The beaker below is filled with symbols of all kinds of elements, both natural and synthetic. Cross out the symbols of the **natural** elements to find the hidden letter. Then list the synthetic elements and their symbols. The hidden letter is _____ .

Es	Sg	Bh	Hs	Lr	Tc	Cm	Mt
Np	Fm	No	Rf	Bk	Md	Hs	Cf
Am	La	Ce	Lr	Es	Os	Pr	Am
Mt	Ga	Al	Ge	Cf	Sc	As	Cm
Bk	Se	Dy	Sb	Po	Li	Te	No
Hs	At	Te	Mo	Ar	Mn	Ne	Es
Bh	He	Ni	Cf	Fe	Zn	Ga	Cm
Sg	Ag	Rh	Tc	Es	Sr	Pt	Cf
Rf	Cm	Fm	No	Bk	Sg	Es	Db
	Hs	Tc	Bh	Np	Am	Md	

ELEMENT	SYMBOL	ELEMENT	SYMBOL	ELEMENT	SYMBOL

Name ______________________________ Date ____________

Activity 48: **Transuranium Elements Match**

Directions: Match the names of the seventeen numbered elements with the lettered hints about the source of their names. Write the correct letter on the line next to the element name.

ANSWER	ELEMENT	HINTS
______	1. Americium	(a) American nuclear chemist
______	2. Berkelium	(b) French chemist
______	3. Bohrium	(c) Danish physicist
______	4. Californium	(d) A coveted prize
______	5. Curium	(e) German province
______	6. Dubnium	(f) Eighth planet from the sun
______	7. Einsteinium	(g) The country you are living in
______	8. Fermium	(h) Town in California
______	9. Hassium	(i) Russian city
______	10. Lawrencium	(j) A man's first name
______	11. Meitnerium	(k) Gold-foil experiment
______	12. Mendelevium	(l) West-coast state
______	13. Neptunium	(m) Austrian physicist
______	14. Nobelium	(n) First published periodic table
______	15. Plutonium	(o) Italian nuclear scientist
______	16. Rutherfordium	(p) Developer of the theory of relativity
______	17. Seaborgium	(q) Ninth planet from the sun

The Newest Elements

Preliminary Handout:

- The Newest Elements (reproducible reading)

Activities:

49. Newest Elements Quiz (reproducible)
50. Periodic Table Update (Internet activity; teacher guide page)

The Newest Elements

There are currently six elements that have been made since 1990, which have not been officially named. These elements currently have Latin names and three-letter symbols.

- Ununnilium (Uun) has an atomic number of 110 and belongs to group 10 of the periodic table. It is made by fusing lead-208 and nickel. Depending on the nickel isotope used, the atomic mass for ununnilium is calculated using the fusion equation as 269 or 271. The fusion equation is a method of writing down the elements that are joined together by the fusion reaction. For example, Pb-208 + Ni-61 $\rightarrow$ Uun. By adding the masses of the individual elements (208 + 61), the atomic mass of the new element can be calculated. Ununnilium has a half-life of 270 microseconds.
- Unununium (Uuu) has an atomic number of 111 and a calculated atomic mass of 272. It belongs to group 11. Unununium was first made in Darmstadt, Germany, on December 8, 1994, by fusing bismuth-209 and nickel-64.
- Ununbium (Uub) has an atomic number of 112 and a calculated atomic mass of 277. It belongs to group 12. Ununbium is produced by combining lead-209 and zinc-70. It was made in Darmstadt on February 9, 1996. It has a half-life of 240 microseconds.
- One atom of ununquadium (Uuq) was made in Dubna, Russia, by fusing plutonium-244 and calcium-48. Ununquadium belongs to group 14. It has an atomic number of 114 and a calculated atomic mass of 285. Ununquadium has a 30-second half-life and, although its synthesis has not been confirmed, it is part of the decomposition chain of element Uuo.
- Ununhexium (Uuh) has an atomic number of 116 and a calculated atomic mass of 289. Ununhexium belongs to group 16. Like ununquadium, it is a decomposition product of element Uuo.
- Ununoctium (Uuo) has an atomic number of 118 and a calculated atomic mass of 293. It belongs to group 18. Ununoctium was first made in Berkeley, California, by fusing krypton-86 and lead-208. Only three atoms of ununoctium were identified after scientists spent 11 days searching for signs of its existence. Ununoctium disintegrates in 0.12 milliseconds to form ununhexium.

Elements with atomic numbers 113, 115, and 117 have not been made or identified yet. Scientists continue to make new elements in an effort to produce stable elements—elements which will exist longer than the few seconds or milliseconds of the latest elements. Theoretically, an **island of stability** or group of stable elements exists. If this island of stability can be found, scientists may be able to study the new elements and obtain more information about elemental chemistry.

Name ______________________________ Date ____________

Activity 49: **Newest Element Quiz**

Part A. Directions: Select the correct answer from the choices given.

1. The newest elements are (**natural** or **synthetic**).
2. The newest elements are found in (**period 7** or **period 8**).
3. The symbols for these elements all have (**one**, **two**, or **three**) letters.
4. The newest elements all have (**Latin**, **English**, or **Russian**) names.
5. The (**atomic number** or **atomic mass**) for each of these elements is calculated using the fusion equation.

Part B. Directions: Indicate if the statement is true or false. If **false**, correct the underlined portion to make the statement **true**.

6. Since element number <u>116</u> belongs to group 18, it should have properties similar to the noble gases. ______________________________

7. Since each of these elements has a half-life consisting of a second or less, they <u>are not part</u> of the island of stability.

   ____

8. All of these elements are formed by <u>fission</u> reactions.

   ____

9. All of these elements undergo radioactive decay and emit <u>alpha</u> particles.

   ____

10. The elements with atomic numbers <u>111, 112, and 114</u> are transition elements.

    ____

11. As of June 1999, the elements with atomic numbers <u>113, 115, and 117</u> have not been synthesized or identified.

    ____

12. Only one atom of <u>element Uub</u> was made in Dubna, Russia.

    ____

Activity 50: **Periodic Table Update**

Objective: Students will place the new elements in the correct places on the periodic table and predict possible physical and chemical properties based on group number.

Materials:

- Copy of the periodic table that does not include the newest elements (or a blank periodic table format)
- Access to Internet for latest data on elements (see Appendix B)

Procedure:

1. Have students add the atomic number, calculated atomic mass (written in brackets), and three-letter symbol for the newest elements to their copies of the periodic table.
2. Divide students into six groups. Assign a different element to each of the groups.
3. Each group will predict the possible physical and chemical properties for their element based on its location on the periodic table. Each group will submit the following information for its element.
   (a) Element name and symbol
   (b) Atomic number
   (c) Atomic mass [calculated]
   (d) Number of protons, neutrons, and electrons
   (e) Electron configuration
   (f) Number of valence electrons
   (g) Group number
   (h) Fusion reaction used to form element
   (i) Physical properties (phase at room temperature, melting point, boiling point)
   (j) Chemical properties (reactivity, electronegativity, possible ions, oxidation states)
4. Physical and chemical properties are to be predicted based on data for other elements in the same group. Some of the properties may not be predictable if there is no pattern seen in the group.
5. Internet access can be used to verify predictions or obtain information.

Results: Students will use knowledge of the periodic table to describe the newest elements and predict their properties.

Extensions:

1. Have students predict information for elements 113, 115, and 117.
2. Have students research the theory of the existence of an island of stability.

Answer Key

Answer Key

Activity 4: **The Metals Crossword**

ACROSS
1. Cobalt (Co)
3. Silver (Ag)
6. Titanium (Ti)
7. Sodium (Na)
9. Nickel (Ni)
13. Magnesium (Mg)
15. Iron (Fe)
16. Zinc (Zn)
17. Calcium (Ca)
19. Lead (Pb)

DOWN
2. Aluminum (Al)
4. Platinum (Pt)
5. Potassium (K)
8. Tin (Sn)
10. Chromium (Cr)
11. Lithium (Li)
12. Mercury (Hg)
14. Gold (Au)
17. Copper (Cu)
18. Iridium (Ir)

Activity 5: **Nonmetals Word Find**

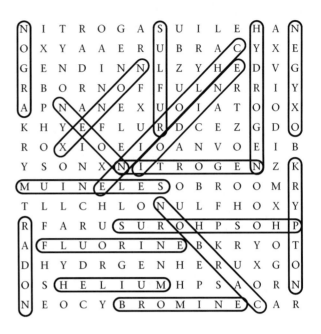

ELEMENT	SYMBOL
Hydrogen	H
Helium	He
Carbon	C
Nitrogen	N
Oxygen	O
Fluorine	F
Neon	Ne
Phosphorus	P
Sulfur	S
Chlorine	Cl
Argon	Ar
Selenium	Se
Bromine	Br
Krypton	Kr
Iodine	I
Xenon	Xe
Radon	Rn

N	I	T	R	O	G	A	S	U	I	L	E	H	A	N
O	X	Y	A	A	E	R	U	B	R	A	C	Y	X	E
G	E	N	D	I	N	N	L	Z	Y	H	E	D	V	G
R	B	O	R	N	O	F	F	U	L	N	R	R	I	Y
A	P	N	A	N	E	X	U	O	I	A	T	O	O	X
K	H	Y	E	F	L	U	R	D	C	E	Z	G	D	O
R	O	X	I	O	E	I	O	A	N	V	O	E	I	B
Y	S	O	N	X	N	I	T	R	O	G	E	N	Z	K
M	U	I	N	E	L	E	S	O	B	R	O	O	M	R
T	L	L	C	H	L	O	N	U	L	F	H	O	X	Y
R	F	A	R	U	S	U	R	O	H	P	S	O	H	P
A	F	L	U	O	R	I	N	E	B	K	R	Y	O	T
D	H	Y	D	R	G	E	N	H	E	R	U	X	G	O
O	S	H	E	L	I	U	M	H	P	S	A	O	R	N
N	E	O	C	Y	B	R	O	M	I	N	E	C	A	R

Activity 6: **Mixed-Up Table**

Aluminum 26.982	Boron 5	Gallium Ga	Nitrogen 14.007	Silicon 28.086
Arsenic As	Bromine Br	Germanium 32	Oxygen 15.999	Sulfur 32.066
Antimony Sb	Carbon 12.011	Indium 49	Phosphorus 30.974	Tellurium 52
Astatine (210)	Chlorine Cl	Iodine 53	Polonium 84	Thallium 81
Bismuth 83	Fluorine 18.998	Lead Pb	Selenium Se	Tin Sn

Corrected Table

Boron	Carbon	Nitrogen	Oxygen	Fluorine
Aluminum	**Silicon**	Phosphorus	Sulfur	Chlorine
Gallium	**Germanium**	**Arsenic**	Selenium	Bromine
Indium	Tin	**Antimony**	**Tellurium**	Iodine
Thallium	Lead	Bismuth	Polonium	**Astatine**

Antimony, arsenic, astatine, boron, germanium, silicon, and tellurium are the **semimetals.**

Activity 8: **Phases Puzzle**

Liquids		Gases		Gases	
ELEMENT	SYMBOL	ELEMENT	SYMBOL	ELEMENT	SYMBOL
Gallium	Ga	Hydrogen	H	Chlorine	Cl
Bromine	Br	Helium	He	Argon	Ar
Cesium	Cs	Nitrogen	N	Krypton	Kr
Mercury	Hg	Oxygen	O	Xenon	Xe
		Fluorine	F	Radon	Rn
		Neon	Ne		

The letters formed by the symbols are *L* and *G*.

Tl	At	Bi	Se	Sn	Te	Ac	
Yb	Sr	Zn	Mn	Al	Li	Ra	Ce
Fm	**(Hg)**	Mg	Fr	S	B	Ti	Er
Nd	**(Br)**	C	Fe	Ca	K	V	Md
No	**(Cs)**	**(Ga)**	Hf	Si	Ge	Ta	Tm
Es	I	Cr	Mt	Ba	Rb	Zn	Pr
Cf	Cu	Sc	**(O)**	**(He)**	Na	Lr	U
Bk	Rf	**(Cl)**	W	Co	Ni	Tc	Cm
Pm	Rh	**(Ne)**	Pt	Be	Ag	Db	Tb
Pu	Pd	**(F)**	Ir	**(Ar)**	**(H)**	Nb	Am
Dy	Hs	**(Kr)**	Cd	Au	**(N)**	Os	Ho
Th	Re	Zr	**(Rn)**	**(Xe)**	Y	Lu	Np
Pa	Mo	Pb	Po	As	P	Ru	Gd
Sm	Bh	Sb	In	La	Eu	Sg	

Activity 9b: **Mendeleev Quiz**

1. Dmitri
2. 1869
3. mass
4. Russia
5. 101

Activity 11a: **Arrangement of Groups and Periods: Finding Periodicity Patterns**

Possible explanations for the periodicity are given below.

Problem 1: The gazelle population increases rapidly over a seven-year period then suddenly decreases. One possible explanation is that the gazelles have exceeded the carrying capacity for their environment (i.e., there are more gazelles than there is food). As a result, the gazelles starve or do not reproduce causing the population to drop. Once the population is smaller, there is enough food and the population starts to increase again.

Problem 2: There is a quarterly periodicity since there are more appliance trucks leaving at the end of March, June, September, and December. This probably ties into the quarterly production reports found in most manufacturing firms.

There is also a yearly periodicity since there are even more trucks near the end of December. This is probably due to the end of the year push for the annual production report.

Activity 11b: **Finding Number and Shape Patterns**

1. □ □ □ (The pattern is an increasing number of triangles and boxes)
2. 3 (Pattern is ÷3, +4, ÷3, +4, ÷3)
3. $\boxed{3}$ (The pattern is an increasing number of angles followed by a box with number of angles written inside)
4. 33 (The pattern is -1, +2, -3, +4, -5, +6, -7, +8, -9)
5. ○ (The pattern is 2 circles, 2 triangles, 1 circle above, 1 triangle above, 2 circles, 2 triangles, 1 circle below)
6. ⬡ (The pattern is a progression of figures with increasing number of sides)
7. 314 (The pattern is 2, 2^2, 2^2-1, 3^2, 3^2-4, 5, 5^2, 5^2–7, 18^2, 18^2–10)
8. △ (The pattern is triangle, arrow, and circle with arrow; the triangle always points in the opposite direction from the arrows)
9. 71 (The pattern is -8, -4, +6, -8, -4, +6, etc.)
10. 59 (The pattern is +4, +4^2, -5, -(5^2), +6, +6^2, -7, -(7^2), +8, +8^2)

Activity 12: **An Elemental Trip Through Europe**

(ThIS) past summer we took a long (VAcAtION) through Europe. We saw many (PAlAcEs) during our trip. (SiNCe) Europe is (WHeRe) much of (AmErICa'S) cultural (HErITaGe) is derived, we tried to see as many historical (PLaCeS) as possible.

We spent (NiNe) days in (FRaNCe). In (PArIS), after visiting the Eiffel Tower and the (ArC) de Triomphe, we (SAt) at a (CaFe) to watch the people. We saw the cathedral at Chartres on our drive to the (SOUTh) of (FRaNCe). We took a short side trip (OVEr) the Pyrenees to Spain to visit Barcelona and (OBTaIn) a flavor of Castilian and Moorish influences. Then (BaCK) to (SOUThErN) (FRaNCe) and the beautiful (BeAcHeS) of (NiCe).

After seeing the French Riviera, we drove to Italy and through the (WINe) country to Pisa. From Pisa it was a short drive following the (ArNo) River to Florence and the (BIrThPLaCe) of Renaissance art. Our next stop was the Eternal City, Rome. We had to see the Coliseum, the (VAtICaN) and the Pope. From Rome we drove (SOUTh) along the Italian coastline to (NaPoLi) and Pompeii. Then we headed northeast (SInCe) we could not leave Italy without a gondola ride on the canals of Venice.

From sea level at Venice, we headed for (VErY) high ground as we drove (OVEr) the (AlPS) to Switzerland. We had (HoPEs) of skiing, down the (AlPINe) slopes and (AlSO) of buying a (FAmOUS) (SWISS) WAtCH). However, there was not enough (SnOW) for skiing so we headed for Austria and the (SiTeS) from the movie *The (SOUNd) of Music.*

From Vienna and Salzburg and beautiful waltzes, we went to Germany to see the (CaSTlEs) of the (BLaCK) Forest and (HeAr) polkas. We survived the autobahn and visited the Peace (GaTe) in (BErLiN).

From (BErLiN) we headed (BAcK) toward (FRaNCe) to (CrOsS) the English Channel. We decided to (CrOSS) by ferry so we could see the (WHITe) (ClIFFS) of Dover. We had (NiNe) days left to see (As) much (OF) England as possible. We, of course, saw Buckingham (PaLaCe) and the guards (WITh) their tall fuzzy (BLaCK) hats and stern (FAcEs).

We caught a (NeW) (PLaY) at Covent Gardens and a Shakespearean (PLaY) at Stratford-on-Avon. We drove as (FAr) north as Sherwood Forest and Nottingham to relive the (FAmOUS) tales of Robin Hood. Then (SOUTh) and west to see Stonehenge on the Salisbury (PLaInS) and the (BeAcHeS) at Bournemouth and (SOUThAmPtON).

(BY) then it (WAs) time to return to London's Heathrow Airport for the (PLaNe) flight (BaCK) to (AmErICa). (WHAt) an amazing (VAcAtION)! We will (NeVEr) forget our elemental trip through Europe.

Activity 13: **Elemental Math**

1. Cl + He = 17 + 2 = 19, answer is K.
2. Tc + Ag - Ne = 43 + 47 - 10 = 80, answer is Hg.
3. (H + Br) ÷ Li = (1 + 35) ÷ 3 = 12, answer is Mg.
4. (Cs + Pa) ÷ He = (55 + 91) ÷ 2 = 73, answer is Ta.
5. Na × Be + C = 11 × 4 + 6 = 50, answer is Sn.
6. Re ÷ (Sc + Be) = 75 ÷ (21 + 4) = 3, answer is Li.
7. Fm ÷ Mn × Be = 100 ÷ 25 × 4 = 16, answer is S.
8. In ÷ N × B + Es - Mn = 49 ÷ 7 × 5 + 99 – 25 = 109, answer is Mt.
9. C^{He} × Ne ÷ Mg = 6^2 × 10 ÷ 12 = 30, answer is Zn.
10. Te + Xe – Sg + As + Pd = 52 + 54 – 106 + 33 + 46 = 79, answer is Au.
11. (Pm - Sb) × O + F = (61 - 51) × 8 + 9 = 89, answer is Ac.
12. [(Zr + Ge) ÷ F] + [Xe ÷ C × Li - Mg] = [(40 + 32) ÷ 9] + [54 ÷ 6 × 3 - 12] = 23, answer is V.

Activity 14: **Cooking with the Elements**

1. For breakfast we (FrY) eggs, (BaCoN) and (HAsH) (BrOWN) potatoes, and toast (WHeAt) or (WHITe) bread. Or, we can have (PaNCaKEs) or waffles and sausage, or (CeReAl), such as (CoRn) (FLaKEs) or (RaISiN) (BRaN), with milk.
2. (HeAlThY) (SnAcKS) would be fruits, such as (BaNaNaS), grapes, (KIWIS), apples, and oranges and different (CHeEsEs) and (CrAcKErS). Of course, most of us would (RaThEr) have (CHIPS), (CoOKIEs), or (CaNDy).
3. For drinks, we (PReFEr) (CoCaCoLa) or another type of soda (OVEr) milk, juice or (WAtEr).
4. Most people have fast food and (USe) the drive (ThRuS) for lunch. They usually have only half an hour and (CHoOSe) (TaCOs) or hamburgers and French (FrIEs). Sometimes they will be (LuCKY) and have a salad, (SOUP), sandwich, or (CHINeSe) take-out. At (OThEr) times, people, especially students, eat (NaCHoS) or (CHILi) cheese (FrIEs).
5. Dinners are the big meals. (ThIS) is (WHeN) families (GaThEr) together after a long day. Dinners usually consist of a main dish containing some type of meat. The meat can be (TbONe), (HAm), pork (CHoPS), chicken, (BaBY) (BaCK) ribs, prime rib, or (FISH).
6. Of course there is always some type of carbohydrate. (ThIS) is usually a potato which we can bake, mash, (FrY), scallop, or boil. For variety, there is also rice or (PAsTa).
7. There usually is a (CHoICe) of vegetables. Some (CHoICeS) are (CoRn), peas, (BrOCCoLi), beans, (AsPArAgUS) or squash.
8. One of my favorite (SOUPS) is (NeW) England (ClAm) chowder. I (SAuTe) the (BaCoN) and (ONiONS) first. Then I add (WAtEr), (ClAmS), celery, and (SPICeS) such as (BaY) leaf, thyme, and marjoram. The diced potatoes and (CReAm) are added about thirty minutes (BeFORe) serving.
9. The best part is dessert. There are many different (CaKEs) and (PIEs). (RhUBaRb) looks like red celery, and is tart (WHeN) baked in (PIEs). Another simple dessert is (ICe) (CReAm). (ThIS) can be served with (OThEr) desserts or (BY) itself, in a (CoNe) or a dish, (WITh) (CHoCoLaTe) or (CaNdY) toppings.
10. A fancy dessert is (CRePEs) with a (BRaNdY) sauce. Many (FINe) restaurants will (LaCe) their desserts with (BRaNDy) to make a flaming dessert.

Activity 15: **Matching Names and Symbols**

Li	1. Lithium
Al	2. Aluminum
Ne	3. Neon
Ca	4. Calcium
S	5. Sulfur
B	6. Boron
Cr	7. Chromium
Zn	8. Zinc
He	9. Helium
P	10. Phosphorus
O	11. Oxygen
Cl	12. Chlorine
Mg	13. Magnesium
Mn	14. Manganese
H	15. Hydrogen
I	16. Iodine
Si	17. Silicon
Ar	18. Argon
C	19. Carbon
N	20. Nitrogen

Activity 16: **Unusual Element Symbols**

SYMBOL	ELEMENT	LATIN NAME
1. Ag	silver	argentum
2. Au	gold	aurum
3. Cu	copper	cuprum
4. Fe	iron	ferrum
5. Hg	mercury	hydrargyrum
6. K	potassium	kalium
7. Na	sodium	natrium
8. Pb	lead	plumbum
9. Sn	tin	stannum
10. Sb	antimony	stibium

Activity 19: **Calculating Protons, Neutrons, and Electrons Given A and Z**

ELEMENT	ELEMENT SYMBOL	ATOMIC NUMBER (Z)	MASS NUMBER (A)	NUMBER OF PROTONS	NUMBER OF NEUTRONS	NUMBER OF ELECTRONS
Carbon	**C**	6	**12**	**6**	6	**6**
Silicon	**Si**	**14**	28	14	**14**	**14**
Iron	Fe	26	56	**26**	**30**	**26**
Gold	**Au**	**79**	**197**	79	118	**79**
Silver	Ag	**47**	**108**	**47**	61	47
Lead	Pb	82	**207**	**82**	125	**82**
Fluorine	**F**	**9**	19	9	**10**	**9**
Oxygen	**O**	8	16	**8**	**8**	**8**
Magnesium	Mg	**12**	**24**	12	12	**12**
Potassium	K	19	**39**	**19**	20	**19**
Copper	**Cu**	**29**	64	**29**	**35**	29
Nitrogen	**N**	7	14	**7**	**7**	**7**
Hydrogen	**H**	1	1	**1**	**0**	**1**
Sodium	Na	**11**	**23**	11	12	**11**

NOTE: Atomic number **(Z)** equals number of protons and electrons in a neutral atom. Mass number **(A)** equals the number of protons and neutrons.

EXTENSION: The table can be modified to add a column to identify neutral atoms, isotopes (which are neutral but which have a different mass number than given on the periodic table), and positive or negative ions.

Activity 20: **Metal Ball Drop**

1. Students should explain the difference in size and depth of each hole.
2. You determine the height from which the ball is dropped by the depth of the hole. The ball dropped from the lowest level makes the shallowest hole. It doesn't have as much energy, and, therefore, does not make as much of an impact in the clay. The ball with the most energy (i.e., the one dropping the farthest) makes the deepest hole.
3. This simulates the electrons because, as electrons drop from different energy levels to the lowest energy level, they give off different amounts of energy. The greater the energy drop, the more energy is given off. The various amounts of energy are seen as the different wavelengths of light in the element spectrum. The wavelengths of light are like the depth of the holes.

Activity 21b: **Ion Charges**

ELEMENT	SYMBOL
Fluorine	**1–**
Magnesium	**2+**
Oxygen	**2–**
Phosphorus	**3–**
Hydrogen	**1+**
Neon	**0**
Chlorine	**1–**
Aluminum	**3+**
Sodium	**1+**
Calcium	**2+**
Lithium	**1+**
Bromine	**1–**
Helium	**0**
Nitrogen	**3–**
Rubidium	**1+**
Barium	**2+**
Potassium	**1+**
Boron	**3+**
Sulfur	**2–**
Iodine	**1–**

Activity 22: **Isotopes and Ions**

ELEMENT	SYMBOL	ATOMIC NUMBER (Z)	MASS NUMBER (A)	NUMBER OF PROTONS	NUMBER OF NEUTRONS	NUMBER OF ELECTRONS	NEUTRAL ISOTOPE, POSITIVE ION NEGATIVE ION
Carbon	**C**	6	12	**6**	**6**	**6**	Neutral
Carbon	C	**6**	**14**	6	8	6	**Isotope**
Oxygen	**O**	8	**16**	**8**	8	10	**2– ion**
Magnesium	Mg	12	24	**12**	**12**	10	**2+ ion**
Uranium	**U**	92	238	**92**	**146**	**92**	Neutral
Uranium	U	**92**	**235**	**92**	143	92	Isotope
Helium	**He**	2	4	**2**	**2**	2	**Neutral**
Sodium	Na	11	**23**	**11**	12	10	**1+ ion**
Chromium	**Cr**	24	52	**24**	**28**	**22**	2+ ion
Chlorine	Cl	17	**35**	**17**	18	**18**	1– ion
Mercury	Hg	80	201	**80**	**121**	**80**	Neutral
Aluminum	**Al**	**13**	**27**	13	14	**10**	3+ ion
Neon	Ne	**10**	20	10	**10**	**10**	Neutral
Hydrogen	**H**	1	2	**1**	**1**	**1**	Isotope
Hydrogen	**H**	**1**	**1**	1	0	**0**	1+ ion
Phosphorus	**P**	**15**	**31**	15	16	15	**Neutral**
Lithium	Li	**3**	**7**	**3**	4	3	Neutral
Potassium	**K**	**19**	**39**	19	20	19	**Neutral**
Iron	Fe	26	**56**	**26**	30	23	**3+ ion**

Activity 27: **Periodic Riddles**

1. hydrogen
2. fluorine
3. potassium
4. gold
5. radium
6. nitrogen
7. tin
8. iron
9. titanium
10. silicon

Activity 29: **Electron Configuration—Element to Electron Configuration**

1. Li = $1s^2 2s^1$
2. Mg = $1s^2 2s^2 2p^6 3s^2$
3. O = $1s^2 2s^2 2p^4$
4. Ar = $1s^2 2s^2 2p^6 3s^2 3p^6$
5. Fe = $1s^2 2s^2 2p^6 3s^2 3p^6 4s^2 3d^6$
6. Br = $1s^2 2s^2 2p^6 3s^2 3p^6 4s^2 3d^{10} 4p^5$
7. Hg = $1s^2 2s^2 2p^6 3s^2 3p^6 4s^2 3d^{10} 4p^6 5s^2$ $4d^{10} 5p^6 6s^2 4f^{14} 5d^{10}$
8. Si = $1s^2 2s^2 2p^6 3s^2 3p^2$
9. U = $1s^2 2s^2 2p^6 3s^2 3p^6 4s^2 3d^{10} 4p^6 5s^2$ $4d^{10} 5p^6 6s^2 4f^{14} 5d^{10} 6p^6 7s^2 5f^3 6d^1$
10. K = $1s^2 2s^2 2p^6 3s^2 3p^6 4s^1$

Activity 30: **Electron Configuration—Electron Configuration to Element Name**

1. $1s^1$ = Hydrogen
2. $1s^22s^22p^1$ = Boron
3. $1s^22s^22p^5$ = Fluorine
4. $1s^22s^22p^63s^1$ = Sodium
5. $1s^22s^22p^63s^23p^3$ = Phosphorus
6. $1s^22s^22p^63s^23p^64s^23d^7$ = Cobalt
7. $1s^22s^22p^63s^23p^64s^23d^{10}4p^6$ = Krypton
8. $1s^22s^22p^63s^23p^64s^23d^{10}4p^65s^24d^{10}$ = Cadmium
9. $1s^22s^22p^63s^23p^64s^23d^{10}4p^65s^24d^{10}5p^66s^2$ $4f^{14}5d^{10}6p^2$ = Lead
10. $1s^22s^22p^63s^23p^64s^23d^{10}4p^65s^24d^{10}5p^66s^2$ $4f^{14}5d^{10}6p^67s^25f^{11}$ = Einsteinium

Activity 32b: **Group Codes**

ELEMENT	GROUP NUMBER
Potassium	1
Fluorine	17
Silicon	14
Radon	18
Gold	11
Beryllium	2
Oxygen	16
Neon	18
Rubidium	1
Tungsten	6
Arsenic	15
Calcium	2
Sodium	1
Aluminum	13
Chromium	6
Iodine	17
Helium	18
Nitrogen	15
Iron	8
Zinc	12

Activity 33: **Groups 4–11 Crossword**

ACROSS

3. Vanadium (Group 5)
5. Bohrium (Group 7)
7. Osmium (Group 8)
9. Tantalum (Group 5)
11. Meitnerium (Group 9)
13. Rhenium (Group 7)
15. Manganese (Group 7)
16. Cobalt (Group 9)
17. Titanium (Group 4)
19. Nickel (Group 10)

DOWN

1. Palladium (Group 10)
2. Molybdenum (Group 6)
4. Platinum (Group 10)
6. Hafnium (Group 4)
8. Silver (Group 11)
10. Technetium (Group 7)
12. Tungsten (Group 6)
14. Niobium (Group 5)
16. Copper (Group 11)
18. Iron (Group 8)

Activity 34: **Groups 12–16 Word Game**

1. SILICON, Group 14.
2. NITROGEN, Group 15.
3. OXYGEN, Group 16.
4. BORON, Group 13.
5. MERCURY, Group 12.
6. GALLIUM, Group 13.
7. CARBON, Group 14.
8. SULFUR, Group 16.
9. ALUMINUM, Group 13.
10. ZINC, Group 12.
11. ARSENIC, Group 15.
12. SELENIUM, Group 16 .
13. CADMIUM, Group 12.
14. TIN, Group 14.
15. POLONIUM, Group 16.
16. BISMUTH, Group 15.
17. PHOSPHORUS, Group 15.
18. CARBON,Group 14.
19. TIN, Group 14.
20. INDIUM, Group 13.

Activity 35: **Alkali Metal Clues**

ELEMENT	SYMBOL	ATOMIC NUMBER
Lithium	Li	3
Sodium	Na	11
Potassium	K	19
Rubidium	Rb	37
Cesium	Cs	55
Francium	Fr	87

1. LITHIUM
2. FRANCIUM
3. POTASSIUM
4. CESIUM
5. SODIUM
6. RUBIDIUM

Activity 36: **Alkali Metals Quiz**

1. Group 1.
2. one, 1+.
3. cation.
4. lithium, Li.
5. francium, Fr.
6. Cesium, francium.
7. (lithium, sodium, potassium, rubidium), knife.
8. oxygen.
9. hydrogen.
10. bases.

Activity 37: **Alkaline Earth Metals: Sources and Uses**

ELEMENT	SOURCE	USES
Strontium	Strontianite	Special glass for TVs
Magnesium	Seawater	Sacrificial electrodes
Beryllium	Beryl	Spark-proof tools
Calcium	Dolomite	Manufacture of rare earth metals
Barium	Barite	Drilling fluids and body imaging
Strontium	Celestite	Red flares and fireworks
Calcium	Calcite	Drying agent and cement
Magnesium	Dolomite	Laxatives, dyes and Epsom salts
Radium	Pitchblende	Cancer treatment and watch dials

Activity 38: **Alkaline Earth Metals Puzzle**

All alkaline earth metals form ions with a <u>2+</u> charge.

ELEMENT	SYMBOL	ATOMIC NUMBER
Beryllium	Be	4
Magnesium	Mg	12
Calcium	Ca	20
Strontium	Sr	38
Barium	Ba	56
Radium	Ra	88

(continued)

Activity 38 *(continued)*

Bi	Co	18	Sc	Ni	43	82	Rb	Cs	87
Md	24	**Be**	**Mg**	Mo	42	41	**56**	Ge	79
Cu	**88**	Mn	29	**4**	Na	6	**20**	15	84
36	Mo	Fe	Cu	**Ca**	Ru	**Ca**	**Ra**	**Ba**	Xe
85	Sc	17	As	**Ba**	77	Tl	**88**	57	55
52	34	Au	**Mg**	37	76	Hg	**12**	10	Ta
Fe	He	**Sr**	Ag	51	82	19	13	Pb	Re
Hf	**4**	**38**	**20**	**Sr**	Sb	Co	Ga	Li	Pt
Rn	21	3	Ti	Sn	Pb	Eu	16	34	Pd
22	Ti	5	Zr	Ti	Ta	Sm	40	54	Kr

Activity 39: **Halogen Color Codes**

ELEMENT	SYMBOL	ATOMIC NUMBER
Fluorine (yellow)	F	9
Chlorine (green)	Cl	17
Bromine (red)	Br	35
Iodine (violet)	I	53
Astatine (black)	At	85

1. Bromine
2. Iodine
3. Astatine
4. Fluorine
5. Bromine
6. Chlorine
7. Astatine
8. Chlorine
9. Iodine
10. Fluorine

Activity 40: **Halogens Quiz**

Part A

1. Group 17
2. anions, 1-
3. nonmetals
4. fluorine
5. acids

Part B

6. True
7. False, corrosive
8. True

Activity 41: **Noble Gases Puzzle**

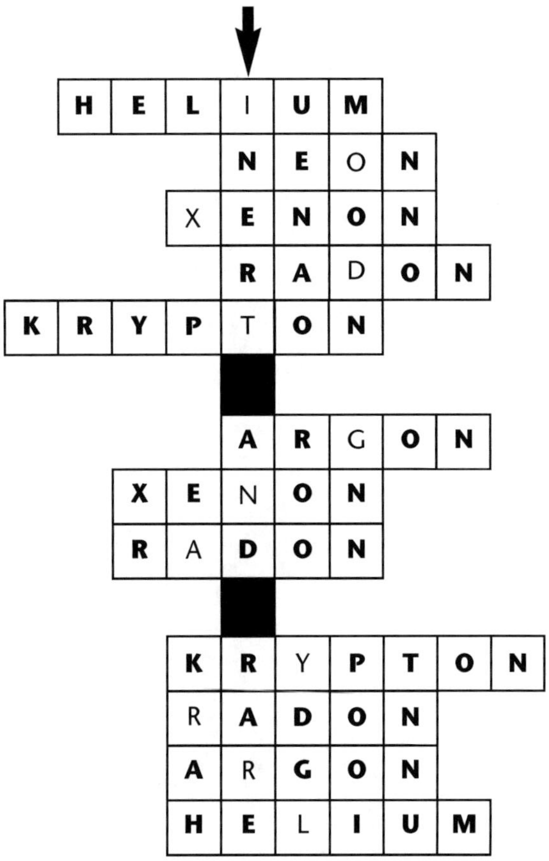

Activity 42: **Noble Gas Pictionary®**

ELEMENT	SYMBOL	ATOMIC NUMBER
Helium	He	2
Neon	Ne	10
Argon	Ar	18
Krypton	Kr	36
Xenon	Xe	54
Radon	Rn	86

1. Neon
2. Radon
3. Helium
4. Xenon
5. Krypton
6. Argon

Activity 43: **Transition Elements Crossword**

ACROSS
1. Cobalt (Co)
6. Tungsten (W)
7. Rhodium (Rh)
8. Silver (Ag)
11. Chromium (Cr)
14. Bohrium (Bh)
16. Erbium (Er)
18. Fermium (Fm)
19. Yttrium (Y)
20. Dysprosium (Dy)
21. Nickel (Ni)

DOWN
1. Copper (Cu)
2. Lanthanum (La)
3. Osmium (Os)
4. Rhenium (Re)
5. Gold (Au)
9. Technetium (Tc)
10. Holmium (Ho)
12. Hafnium (Hf)
13. Iron (Fe)
15. Iridium (Ir)
17. Molybdenum (Mo)

Activity 44: **Transition Element Group Numbers**

ELEMENT	GROUP	ELEMENT	GROUP
Bohrium	7	Osmium	8
Cadmium	12	Palladium	10
Cobalt	9	Platinum	10
Copper	11	Rhenium	7
Chromium	6	Rhodium	9
Dubnium	5	Ruthenium	8
Gold	11	Rutherfordium	4
Hafnium	4	Scandium	3
Hassium	8	Seaborgium	6
Iridium	9	Silver	11
Iron	8	Tantalum	5
Lawrencium	3	Technetium	7
Lutetium	3	Titanium	4
Manganese	7	Tungsten	6
Meitnerium	9	Vanadium	5
Mercury	12	Yttrium	3
Molybdenum	6	Zinc	12
Nickel	10	Zirconium	4
Niobium	5		

Activity 45: **Find the Series**

Cf	Ge	Ni	As	Sb	Sn	Sg
At	<u>Er</u>	Xe	Au	Hf	Re	Ac
Te	<u>Yb</u>	Ag	Si	Br	Ar	No
Cl	<u>Tm</u>	Mg	Nb	Mo	Os	Es
Br	<u>La</u>	Ta	He	Rh	Ti	Am
Kr	<u>Ce</u>	Cs	Ne	Ta	Ir	Zr
Rn	<u>Pr</u>	Pb	Bi	Po	Br	Ga
Xe	<u>Nd</u>	Li	Na	Mn	Ca	Be
Pt	<u>Pm</u>	Ra	Lr	Lu	Tc	Th
Rb	<u>Sm</u>	Cu	Sc	Cd	Hg	Pu
Sr	<u>Eu</u>	Co	Ba	Zn	Ru	Np
Pa	<u>Gd</u>	<u>Tb</u>	<u>Dy</u>	<u>Ho</u>	<u>Lu</u>	Si
Cr	Tl	Nb	Pd	Bk	Md	Al

(continued)

Activity 45 *(continued)*

This series of elements is between the atomic numbers of 57 and 71: the lanthanide series.

ELEMENT	SYMBOL
Cerium	Ce
Dysprosium	Dy
Erbium	Er
Europium	Eu
Gadolinium	Gd
Holmium	Ho
Lanthanum	La
Lutetium	Lu
Neodymium	Nd
Praseodymium	Pr
Promethium	Pm
Samarium	Sm
Terbium	Tb
Thulium	Tm
Ytterbium	Yb

Activity 46: **Actinide Series Word Find**

Actinium	Einsteinium	Nobelium
Americium	Fermium	Plutonium
Berkelium	Lawrencium	Protactinium
Californium	Mendelevium	Thorium
Curium	Neptunium	Uranium

P	O	N	E	P	T	U	N	I	U	M	P	P	L	K	J	H	G	F
M	N	B	V	C	X	Z	A	S	U	Y	T	R	Q	W	E	R	T	Y
Q	W	E	F	E	R	M	J	I	U	M	P	O	J	H	G	H	D	S
A	Q	Z	B	E	F	C	N	A	R	O	H	T	H	O	R	I	U	M
P	C	A	L	I	F	O	R	N	I	U	M	A	A	A	C	V	B	N
Y	U	R	A	D	T	D	A	G	O	U	R	C	S	E	M	Y	H	N
X	R	E	M	U	I	U	N	E	I	O	G	T	S	S	U	I	M	M
N	I	W	L	M	E	N	D	V	A	N	X	I	L	U	I	A	U	L
O	U	P	B	E	R	K	E	L	I	U	M	N	O	B	C	I	I	O
B	M	V	R	A	G	L	A	W	R	C	Q	I	Q	Y	N	I	N	X
E	I	N	S	T	E	I	N	I	U	M	R	U	C	I	E	M	A	Z
L	Z	X	Y	D	P	L	U	T	O	U	I	M	T	B	R	C	R	A
I	P	O	N	C	A	L	I	F	O	I	D	C	U	R	W	G	U	I
U	H	E	M	U	I	C	I	R	E	M	A	I	L	R	A	H	G	F
M	M	Q	U	E	E	T	Y	U	I	R	P	O	I	W	L	T	R	E
Z	X	C	V	B	N	M	L	K	J	E	Q	W	E	L	T	Y	U	I
T	Y	U	R	W	A	L	K	J	H	F	Z	X	C	V	B	N	M	L

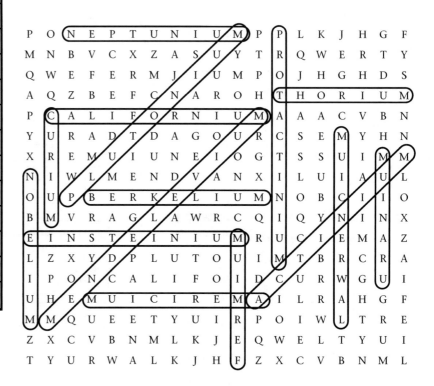

Activity 47: **Synthetic Elements Puzzle**

The hidden letter is N.

Es	Sg	Bh	Hs	Lr	Tc	Cm	Mt
Np	Fm	No	Rf	Bk	Md	Hs	Cf
Am	**La**	**Ce**	Lr	Es	**Os**	**Pr**	Am
Mt	**Ga**	**Al**	**Ge**	Cf	**Sc**	**As**	Cm
Bk	**Se**	**Dy**	**Sb**	**Po**	**Li**	**Te**	No
Hs	**At**	**Te**	**Mo**	**Ar**	**Mn**	**Ne**	Es
Bh	**He**	**Ni**	Cf	**Fe**	**Zn**	**Ga**	Cm
Sg	**Ag**	**Rh**	Tc	Es	**Sr**	**Pt**	Cf
Rf	Cm	Fm	No	Bk	Sg	Es	Db
	Hs	Tc	Bh	Np	Am	Md	

ELEMENT	SYMBOL	ELEMENT	SYMBOL
Americium	Am	Lawrencium	Lr
Berkelium	Bk	Meitnerium	Mt
Bohrium	Bh	Mendelevium	Md
Californium	Cf	Neptunium	Np
Curium	Cm	Nobelium	No
Dubnium	Db	Rutherfordium	Rf
Einsteinium	Es	Seaborgium	Sg
Fermium	Fm	Technetium	Tc
Hassium	Hs		

Activity 48: **Transuranium Elements**

1. G
2. H
3. C
4. L
5. B
6. I
7. P
8. O
9. E
10. J
11. M
12. N
13. F
14. D
15. Q
16. K
17. A

Activity 49: **Newest Elements Quiz**

Part A

1. synthetic
2. Period 7
3. three
4. Latin
5. atomic mass

Part B

6. False, 118
7. True
8. False, fusion
9. True
10. False, 110 and 111
11. True
12. False, Uuq

Appendices

Name ______________________________ Date ____________

Bonus Activity 1: **Elements in the Human Body**

Directions: The word puzzle at the bottom of this page contains several elements that are all found in the human body. Find the names of these elements using the list of symbols in the Element Symbols Box as clues. (Not all symbols are used.) Refer to the periodic table to find the element names. Some letter clues are included to help you. (*Example:* The symbol for oxygen is O, so you would write [O][X][Y][G][E][N] in the spaces provided.) After completing the puzzle, use the vertical column under the arrow to answer the following question:

What does the human body need in order to be healthy?

____

C	H	Na
Ca	I	O
Cl	K	P
F	Mg	S
Fe	N	Zn

Name ______________________________ Date ____________

Bonus Activity 2: **Elements Found in the Air**

Directions: This puzzle contains seven elements found in the air. Fill in the name of each element in the puzzle below. Some of the letters have been filled in for you. Then write each element, its atomic number and its symbol in the table that follows.
Reading down vertically, from the black arrow, find the answer to the following question:

What is the name of the subatomic particle that has no charge?

____

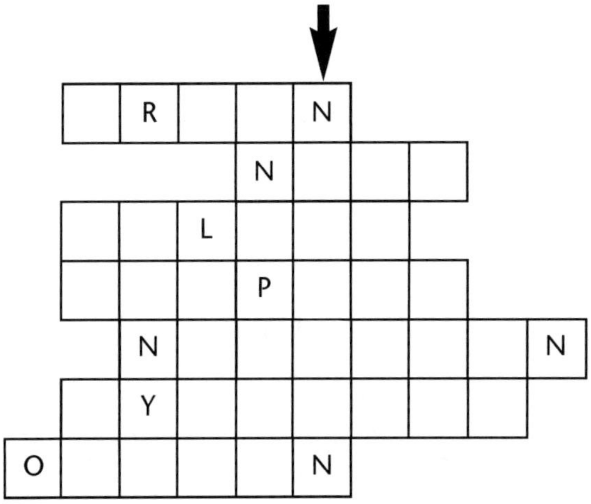

ELEMENT	SYMBOL	ATOMIC NUMBER

Bonus Activity 1: **Elements in the Human Body**

What does the human body need in order to be healthy? Good Nutrition

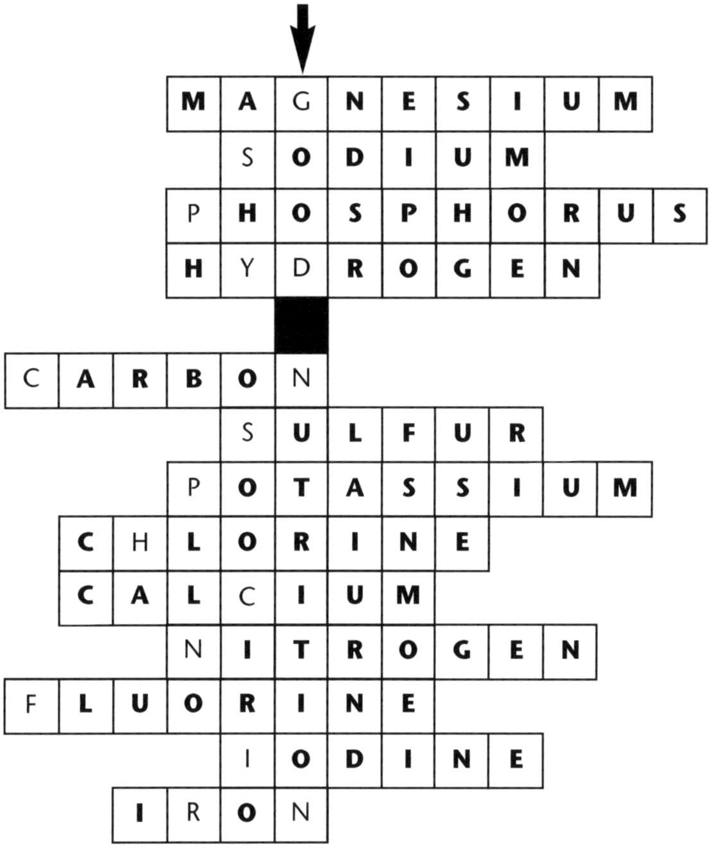

Bonus Activity 2: **Elements Found in the Air**

What is the name of the subatomic particle that has no charge? Neutron

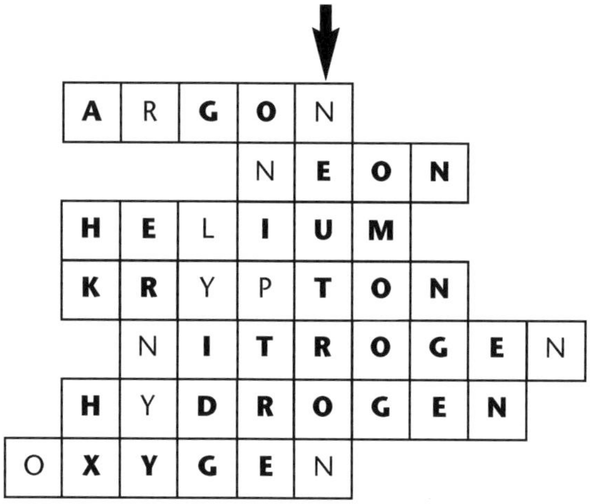

ELEMENT	SYMBOL	ATOMIC NUMBER
Argon	Ar	18
Neon	Ne	10
Helium	He	2
Krypton	Kr	36
Nitrogen	N	7
Hydrogen	H	1
Oxygen	O	8

Appendix B: **Periodic Table Web Sites**

1. http://www.ssc.ntu.edu.sg:8000/chemweb/html/ref-ptable.html
   This web site has links to other web sites with periodic table data.
2. http://chemlab.pc.maricopa.edu/periodic/periodic.html
   This web site gives data on periodic table and also lists other references.
3. http://www.webelements.com/
   This web site provides data on elements and links to other information.
4. http://pearl1.lanl.gov/periodic/
   This is the web site for Los Alamos National Laboratories. It provides data on the elements of the periodic table.
5. http://www.gsi.de/
   This is the web site for Gesellschaft für Schwerionenforschung (GSI), at Darmstadt, Germany. This is one of the companies synthesizing and researching new elements.

Name ______________________________ Date ____________

Appendix C: **Periodic Table (Blank Form)**

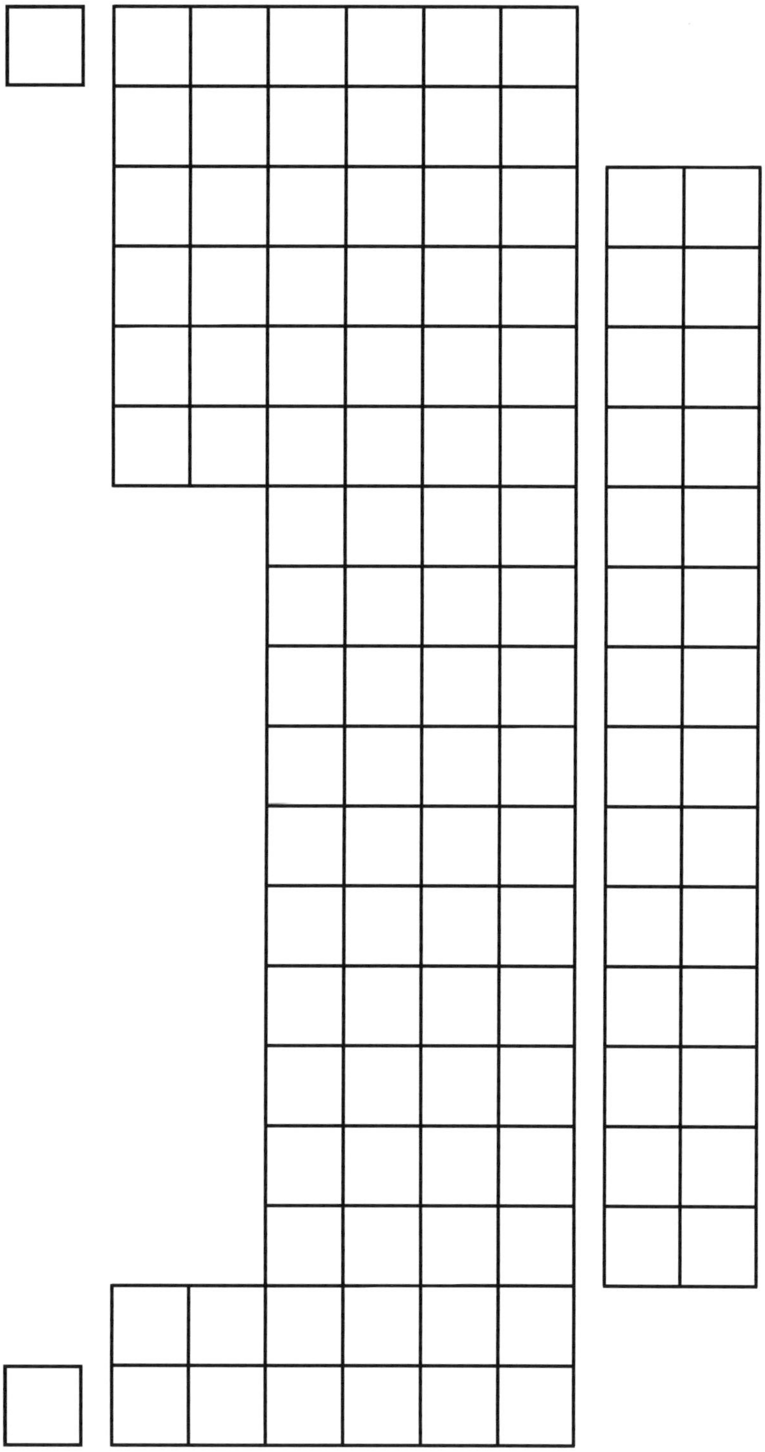